RFID

Design Fundamentals and Applications

RFID
Design Fundamentals and Applications

Albert Lozano-Nieto

CRC Press
Taylor & Francis Group
Boca Raton   London   New York

CRC Press is an imprint of the
Taylor & Francis Group, an **informa** business

CRC Press
Taylor & Francis Group
6000 Broken Sound Parkway NW, Suite 300
Boca Raton, FL 33487-2742

© 2011 by Taylor and Francis Group, LLC
CRC Press is an imprint of Taylor & Francis Group, an Informa business

No claim to original U.S. Government works

Printed in the United States of America on acid-free paper
10 9 8 7 6 5 4 3 2 1

International Standard Book Number: 978-1-4200-9125-0 (Hardback)

Library of Congress Cataloging-in-Publication Data

Lozano-Nieto, Albert.
RFID design fundamentals and applications / author, Albert Lozano-Nieto.
p. cm.
"A CRC title."
Includes bibliographical references and index.
ISBN 978-1-4200-9126-7 (hardcover : alk. paper)
1. Radio frequency identification systems. I. Title.

TK6570.I34L69 2011
006--dc22   2010029742

Visit the Taylor & Francis Web site at
http://www.taylorandfrancis.com

and the CRC Press Web site at
http://www.crcpress.com

To Debbie, for her unconditional and always present love

and support. For having changed my life.

To Ramon Pallas-Areny, for having been a role model. For

instilling in me his passion for education and research.

Contents

Preface

Although the basic concepts of radiofrequency identification (RFID) were developed during World War II, only recently has RFID become a ubiquitous technology in today's industry, market, and society. In 2005, the U.S. Department of Defense required all suppliers to affix passive RFID tags to cases and pallets used for shipment. Almost simultaneously, large private enterprises, such as Wal-Mart, required similar tagging for their own supplies. These two initiatives were instrumental in the acceptance and use of RFID as a tool to increase productivity. More importantly, these mandates have permeated a large number of large and medium-sized businesses that are incorporating RFID techniques to reduce costs and increase revenue. Even small companies are currently considering using this technique in their operations. RFID systems are widely used in the chain supply and logistics industry, for example, to track pallets and cases of products. Large companies benefit from the increased accuracy of their inventory as well as the automation of these processes. Retailers use this technology to prevent theft of their products using tags embedded in the packaging. When the product is purchased by the customer, the cashier or checkout clerk changes the code in the tag, for example, by waving it over a reader or writer, making the tag invisible to the reader located in the exit doors. However, when the product is stolen or the cashier has failed to change the code in the tag, the readers located near the exit doors detect the signal from the tag and start an alarm. RFID is also being used to authenticate that the cold chain in food products has not been compromised, to combat the counterfeit of pharmaceutical drugs, and to establish the authenticity and history of critical parts, for example, in the repair and maintenance of aircraft. New and enhanced applications of using RFID are being developed every day as the number of industries that incorporate this technology expands. This increase in the market size of RFID technologies will, in turn, drive the need for an increasing number of professionals proficient in the evaluation and deployment of these systems.

This book gives the reader an understanding of the fundamental principles involved in the design and characterization of RFID technologies. While the documentation that the manufacturers of each individual component provide is extensive and complete, synthesizing and using this information can represent a challenge to the user: the required information is complex and, more importantly, it is necessary to consult and integrate the information from several manufacturers of different components. This book seeks to solve this problem by presenting a systematic approach in analyzing the different components that make up the whole RFID system, giving the reader the necessary tools to have a clear understanding of them. Readers may choose to read the entire book or just to focus on specific chapters of interest without detriment.

This book contains eight chapters, each one of them a systematic approach to studying the different components that constitute the RFID system. Chapter 1 presents an overview of RFID with a focus on the main components that make up the system. Chapter 2 discusses the types of antennas used in transponders, also known as *tags*. Chapter 3 studies the different parts that make up the transponder, such as its power-harvesting and analog circuits, the different memory structure, and logic circuits. It also discusses the mechanical characteristics of these circuits and their requirements for handling and placement on the transponders. Chapter 4 describes the types of antennas used for RFID interrogators, also known as *readers*, focusing on their design and construction to meet specific requirements. Because of their intrinsic differences, this chapter discusses separately the requirements for antennas based on the frequency of transmission. Chapter 5 discusses the structure and parts that make up the interrogator with a special emphasis on the different types of modulation being used. The chapter also describes the organization and characteristics of commercially available transponders. Chapter 6 studies the different types of communication links between the interrogator and its host computer, either proprietary protocols used by specific manufacturers or standardized communication protocols. Chapter 7 describes the air communications link that is the link between transponders and interrogators. This chapter addresses the different elements of the communication link, the different mode of operation for transponders operating at different frequencies, as well as the principles of arbitration and anticollision. The last chapter in the book, Chapter 8, presents an overview of the different commands used by transponders. This chapter does not pretend to be an exhaustive repository of the available commands for any transponder in the market. Instead, it focuses on the commonalities between the most frequently used functions in the commands employed by different manufacturers. The book finishes with a series of references that can be used by interested readers to explore in depth some of the aspects described in this book.

The majority of books on RFID today are focused on the deployment and the commercial uses of RFID. This book, in turn, is focused on explaining how the different components that make up the RFID system are designed and how they interact with each other. The book is directed toward professionals and students in electronics, telecommunications, and new technologies. However, it will also be useful for those who want to experiment with or enhance their knowledge of RFID systems.

About the Author

Albert Lozano-Nieto is professor of engineering at Penn State. His main teaching responsibilities are focused on the baccalaureate degree in electrical engineering technology at the Wilkes-Barre campus. A native of Barcelona, Spain, he joined Penn State in 1996. Dr. Lozano-Nieto received his doctoral degree in electrical engineering in 1994 and his baccalaureate degree in telecommunication engineering in 1988, both from the Polytechnic University of Catalonia, Barcelona, Spain. He is also an RFID+ certified professional, awarded by the Computing Technology Industry Association (CompTIA), since 2008.

Dr. Lozano-Nieto's research interests focus on the study of errors associated with bioelectrical impedance measurements, as well as on developing new pedagogical approaches to enhance engineering and engineering technology education. On a personal level, he has a strong interest in infrared photography, high-altitude balloons, and amateur rocketry.

1

Basic Principles of Radiofrequency Identification

CONTENTS

Although the widespread use of radiofrequency identification (RFID) is recent, the technology itself is several years old. RFID was first developed during World War II in order to distinguish between friend and enemy aircraft. Several years later, thanks to the technological advances in microelectronics, wireless communications, and computer networks, RFID has come of age. It has now become a technology that is mature enough to be mass-marketed at competitive cost and become a critical player in the global market. This chapter introduces the reader to the different parts that make up a RFID system and provides an overview of the different concepts that will be described in depth in the upcoming chapters.

1.1 Basics of RFID

The basic function of an RFID system is to automatically retrieve the information that has been previously inserted into specific integrated circuits, as seen in Figure 1.1. RFID systems were first developed in automatic identification laboratories as a natural evolution of the different technologies they use. Different automatic identification systems use different methods to transmit the identifying information. For example, bar code technology uses light as the transmission media, while RFID systems use radio waves. In this context, the two main elements of RFID systems are the devices used to carry this information and the equipment used to automatically capture or retrieve the information.

The devices that store and carry the information are called *transponders* or *tags* and are shown in Figure 1.2. Although the industry commonly refers to them as *tags* mainly due to their physical shape and the fact that they are

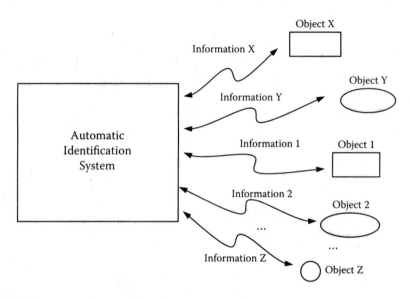

FIGURE 1.1
Information transfer in an automatic identification system.

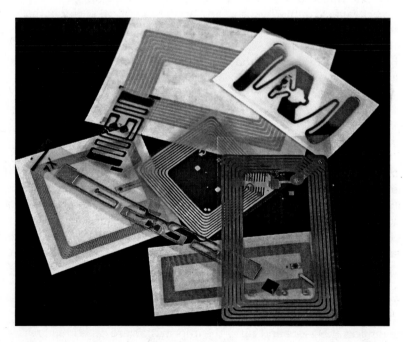

FIGURE 1.2
Several types of transponders used in RFID systems.

mostly used to tag pallets or cases of goods, the name *transponder* reflects better the function of these devices.

The device that is used to capture and transfer information is commonly called a *reader*, because in earlier RFID systems they were only able to read the information sent by the transponders. However, with the development of new functions and applications of RFID systems, the name *interrogator* reflects better the function of this subsystem. Therefore, this book will use the names *transponder* and *interrogator* when referring to these elements.

Transponders must have the circuitry needed to harvest power from the electromagnetic fields generated by the interrogator, the necessary memory elements, as well as the different control circuits. The simplest transponders contain only read-only memory (ROM), while more sophisticated transponders also include random access memory (RAM) and nonvolatile programmable read-only memory (PROM) or electrically erasable programmable read-only memory (EEPROM). ROM usually contains the identification string for the transponder and instructions for its operating system. RAM is normally used for temporary data storage when the transponder communicates with the interrogator. PROM or EEPROM is normally used to incorporate additional functionality depending on the application. The memory capacity of transponders ranges from a single bit to several kilobits. Single-bit transponders are typically used in retail electronic surveillance in which there are only two possible states: *article paid* and *article not paid*. Memory sizes of up to 128 bits are enough to store the individual transponder identification number with several check bits. Memories up to 512 bits are normally user programmable, in which in addition to the individual identification number the memory can hold additional information required by its application. Higher capacity memories can be seen as carriers for the transport of data files. They are also used in applications in which there are several sensors attached to the transponder.

Interrogators have vastly different complexity levels depending on the type of transponders they support as well as their specific purpose. In any case, they all must provide the basic functionality to communicate with the transponders, first by energizing them and second by establishing a communications link. The complexity of the communications link can also vary considerably depending on the desired reliability. The reliability of the communications link between transponder and interrogator can be enhanced by adding parity checks, error detection, and error detection and correction. However, the use of these schemes will result in a lower transmission rate. Figure 1.3 shows a picture of a basic interrogator in which it is possible to see the coiled antenna and some of the electronic circuits.

RFID systems contain more elements in addition to the transponder and interrogator. First, the communication between transponder and interrogator is not possible—or becomes extremely deficient—without the appropriate radiant elements that will transfer the information between these two subsystems in the form of electromagnetic energy. Both transponder and

FIGURE 1.3
Basic RFID interrogator operating in the low-frequency (LF) range.

interrogator need to use the appropriate antennas to transfer information. As will be discussed in Chapter 2, the antenna dimensions in commercially available transponders and interrogators are much shorter than their ideal dimensions. This results in seriously limiting the transfer of energy between interrogators and transponders.

In addition to the transponders, the interrogators, and their antennas, the RFID system requires a *host computer* connected to the interrogator. This host computer provides a certain level of intelligence and acts like the interface between the RFID system and the ultimate application. Therefore, the interrogator must have the means to at least perform basic communication functions with the interrogator. A large number of today's applications require more than one interrogator running on a network rather than the single, stand-alone interrogator that was typical of earlier systems. Finally, an *edge server* is normally used between the host computer and the network in which the application is running.

In addition to these elements (shown in Figure 1.4), most of them hardware based, a typical RFID system contains several software elements. The *firmware* is the software that runs inside the interrogator in order to provide it with its basic functions. Some interrogators may also run one or more software *applications* that enhance the versatility of the interrogator. The name *middleware* encompasses the software applications that run in the background of the system, typically after the host computer. Finally, most applications require the use of one or several *databases* used to link the stored message

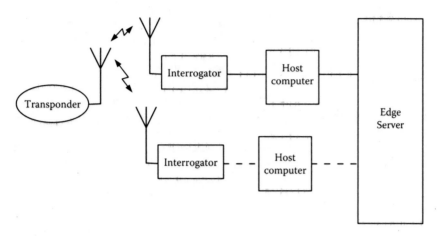

FIGURE 1.4
Basic structure of an RFID system.

sent by the transponder with more detailed information that can identify the transponder and provide additional information.

1.2 Passive versus Active RFID Systems

Passive RFID systems are those systems that use passive transponders. Passive transponders do not have an internal power source. They harvest the energy needed by their internal circuits from the electromagnetic field generated by the interrogator. For this reason, they have a short range, limited to a few feet and often, more realistically, to a few inches. Because they don't have an internal source of energy, the user does not need to worry about the status of the battery. Furthermore, their manufacturing and production costs are very low. The RFID transponders and systems described in this book are all passive.

Active RFID systems, on the other hand, use active transponders. Active transponders have an internal power source, typically a battery that allows broadcasting the signal to the interrogator. Because of not being limited to the power harvested by the antenna, they have an extended read range, typically several hundred feet. However, the inclusion of the power source increases their cost in two ways: the cost of the battery itself as well as the maintenance costs required to check the status of the internal power source and replace it when it has reached an unacceptably low level. The cost of an active transponder is approximately 100 times higher than the cost of a passive transponder. Active transponders typically operate in the ultra-high-frequency (UHF) and microwave ranges. A new type of active RFID transponder is linked to the use of RFID-enabled cell phones that can emulate an active tag. In this case, the problem of battery maintenance is solved as the user keeps the cell phone charged as part of its regular maintenance.

A third type of transponder is called *semipassive* or *battery-assisted transponder*. These transponders also include a battery, but contrary to active transponders, the battery is not used to generate the power to transmit the signal to the interrogator. Instead, the battery is used to support secondary functions like the data logging from different types of sensors. These transponders also harvest the energy from the electromagnetic field generated by the interrogator to power its internal circuits other than the sensing and data-logging parts.

1.3 Functional Classification of RFID Transponders

A functional classification of RFID transponders is based on their electronic product code (EPC) class. EPC is the application of one specific type of RFID technology within the consumer packaged goods industry. Within this classification, RFID transponders are divided into different classes and generations:

Generation 1, Class 0: Passive tags with read-only functionality. These are also called *write one, read many* (WORM) transponders. These transponders are programmed at the factory with their unique identification number. The user is not able to change it or include additional information.

Generation 1, Class 0+: These are also WORM transponders. They differ from Generation 1, Class 0 transponders in that it is the user who programs them. After they have been programmed by the user, no further programming or changing of data is allowed.

Generation 1, Class 1: These transponders were similar to Generation 1, Class 0 or 0+ transponders, but could be read by interrogators from other companies. Gen 1, Class 1 transponders have evolved into the different transponders from Generation 2.

Whereas Generation 1 transponders employ proprietary data structures and can be read only by interrogators manufactured by the same vendor, Generation 2 transponders are vendor neutral in their specifications. This means that as they have developed (following agreed upon standards), they can be read with interrogators from multiple vendors. These transponders provide additional features, for example the elimination of duplicate reads within the read range. They are also more reliable than Generation 1 transponders and support faster read rates. The user of Generation 2 transponders can access specific parts of their memory as these are rewritable transponders.

Generation 2, Class 1: These transponders are WORM transponders. They are also programmed at the factory, but they can be read with equipment from different vendors, support the higher read rates, and have more noise immunity than the Generation 1 transponders.

Generation 2, Class 2: These transponders are rewritable transponders. They can be written several times by the user using equipment different from the vendor's equipment.

Generation 2, Class 3: These transponders are the semipassive or battery-assisted transponders.

Generation 2, Class 4: This refers to active transponders.

Generation 2, Class 5: These transponders are essentially interrogators. Transponders that follow Generation 2, Class 5 must be able to power other transponders.

1.4 Applications and Frequency Selection

RFID systems come in different flavors from the point of view of the frequency they use. Selecting the most adequate frequency is a function of two variables: the technological developments of systems at the different operating frequencies—directly related to the cost of systems—as well as the properties of electromagnetic waves at those different frequencies.

The initial users of RFID systems found systems operating in the low-frequency (LF) range followed a few years later by systems operating in the high-frequency (HF) range. Therefore, their selection of systems was limited to these two broad frequency ranges even if those were not the most adequate for their specific applications. Recent years have seen an increase in the number of available RFID systems operating in the UHF range as well as a reduction of their economic cost. It is possible to expect a similar trend in upcoming years for RFID systems operating in the microwave range to become more available and more affordable. This will help the final users to make decisions on what type of RFID systems to deploy based on technical reasons rather than based on the availability and cost of these systems.

Table 1.1 presents a broad view of the frequency spectrum used by RFID systems as well as the most common frequencies for each band. It is necessary to note that Table 1.1 is not exhaustive because it is possible to find systems, within each frequency band, operating at other frequencies.

The LF band has been used by RFID systems for several years. This fact, together with the less stringent requirements of the electronics operating in this frequency range, has made these systems very cost-effective. Their

TABLE 1.1

Commonly Used Frequency Band for RFID Systems

Frequency Band	Frequency Range	Typical Frequencies Used in RFID Systems
Low Frequency (LF)	100 kHz – 500 kHz	125 kHz 134.2 kHz
High Frequency (HF)	10 MHz – 15 MHz	13.56 MHz
Ultra High Frequency (UHF)	400 MHz – 950 MHz	866 MHz Europe 915 MHz United States
Microwaves (µW)	2.4 GHz – 6.8 GHz	2.45 GHz 3.0 GHz

major drawback is that due to the low frequency of the carrier, these systems can only communicate at low frequency rates. The main advantage of LF RFID systems is that electromagnetic waves operating in the LF range are the least affected by the presence of metals due to the penetration depth of these frequencies. This makes them ideal to be used in applications in which the transponders are surrounded by large metallic masses. Furthermore, LF waves can travel through water with minimal attenuation compared to waves of higher frequencies. This property makes them the frequency range of choice for the identification of animals due to their high water content. Most of these applications are based on implanting these transponders in livestock and pets. Figure 1.5 shows a picture of two different transponders that operate in the LF range in which it is possible to see the multiple loops used by these transponders.

RFID systems operating in the HF range can support higher read rates than LF systems. Although signals in the HF range are more affected by metal and have more attenuation when traveling through water, the transponders used for HF systems can be manufactured at a lower cost than the transponders operating in the LF range. This is due to the fact that the antennas for HF transponders can be made smaller, as shown in the two transponders seen in Figure 1.6. The need for less conductive material to construct the antennas results in a lower cost. Finally, the global uniformity across the world in the frequency of 13.56 MHz makes HF systems able to operate in any country.

While HF RFID systems operate at the single frequency of 13.56 MHz, UHF RFID systems can operate at different frequencies in the UHF band. Moreover, different countries have established different frequency ranges for their UHF RFID operation, making the compatibility between systems more difficult. For example, in North America the assigned band is from 902 MHz to 928 MHz, while in Europe it is from 860 MHz to 868 MHz and in Japan is from 950 MHz to 956 MHz. Other countries and regions have other different allocated frequency ranges. Despite this challenge and due to the developments in microelectronics that allowed the decrease in cost for these

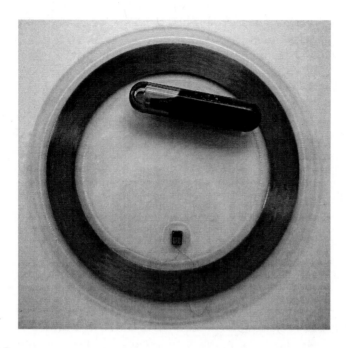

FIGURE 1.5
Low-frequency (LF) RFID transponders require antennas of several hundred feet of coiled wire.

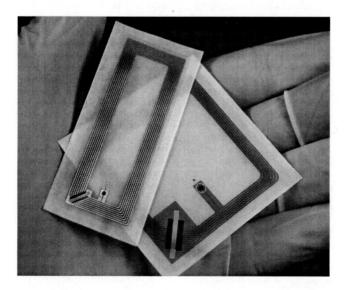

FIGURE 1.6
High-frequency (HF) RFID transponders can use shorter, smaller antennas.

systems, UHF systems have established themselves as a viable alternative to the existing LF and HF systems. One of the main advantages of UHF RFID systems is that their higher carrier frequency allows for a much faster read rate of the transponder information, thus allowing the transmission of higher amounts of data. The communication between transponder and interrogator used in UHF systems is by electrical field instead of the magnetic coupling used by LF and HF systems. The antennas required by UHF systems can also be smaller than the antennas used for LF and HF systems, resulting in higher efficiency. These last two factors contribute to extended read ranges for UHF RFID systems. However, UHF systems also present several drawbacks compared to HF and LF systems. The antennas for transponders operating in the UHF range have very different physical dimensions depending on the parameters that the manufacturer wants to optimize in a given transponder. Figures 1.7 and 1.8 show examples of several UHF transponders in which it is possible to notice these differences.

UHF systems do not work well around metals as these reflect their electromagnetic waves. In addition, water absorbs UHF waves, making them unusable for animal implantation. Furthermore, the extended read range of UHF transponders may become a drawback for those applications that want to keep the read of transponders confined to a certain area.

Although 2.45 GHz is within the UHF range (UHF extends from 300 MHz to 3 GHz), the RFID systems operating in the frequency range extending from 2.45 GHz to 5.8 GHz are commonly called *microwave RFID systems*. In this frequency range, the propagation is limited to line of sight. Microwave signals are also attenuated by water and reflected by metallic objects. The cost of interrogators and transponders used in the microwave range is higher

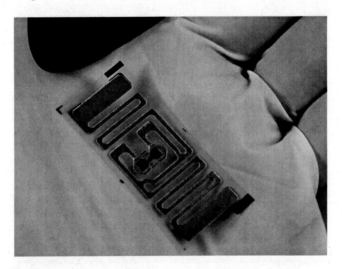

FIGURE 1.7
Ultra-high-frequency (UHF) RFID transponder.

FIGURE 1.8
Several types of antennas used by ultra-high-frequency (UHF) RFID transponders.

than for any other type of system. On the positive side, it is possible to achieve very high read rates, and with the help from spread spectrum techniques, it is possible to achieve very high reliability rates and noise immunity. The higher end of the range, from 5.5 GHz to 6.8 GHz, is still currently under research and development.

2

Antennas for RFID Transponders

CONTENTS

The purpose of the antenna in the transponder of an RFID system is multiple: first, it has to collect power from the electromagnetic field generated by the interrogator. In addition, the antenna must transfer the collected power to its load that is the chip in the transponder in order to turn it on. Finally, the antenna must radiate the data signals generated by the chip back to the interrogator. The antenna should be optimized to minimize energy losses during this process by choosing the type and dimensions of the antenna suited for a specific application and matching its impedance to the impedance of the load. Because these requirements may lead to contradictory solutions, the designer of the system must evaluate them carefully when considering the possible solutions.

This chapter explores the design and performance considerations for antennas in transponders. It starts by presenting a basic review of antenna theory with special emphasis on the differences used for systems in the low-frequency (LF) and high-frequency (HF) ranges compared to the antennas used for ultra-high-frequency (UHF) systems. The chapter continues by describing the types of antennas found in commercial transponders. This is followed by considering the requirements for antenna tuning and antenna matching. The following section describes the different methods for attaching the antenna to the chip on the transponder. The chapter finishes with a review of other antenna factors that affect their performance.

2.1 Review of Basic Antenna Theory for RFID Transponders

2.1.1 Antennas for RFID Transponders Operating in the LH and HF Regions

Electrical current flowing through a conductor generates electromagnetic fields. From the point of view of the concepts described in this chapter, it is possible to distinguish two field regions. The first region is called the *far-field region*. In this region, the generated fields are radiated fields, meaning that the energy propagates through the space with an energy density proportional to the inverse of the distance. The second area is called the *near-field region*. In this area, radiated fields are not prevalent. Instead, attenuating fields, in which the strength of the field decreases with $1/r^3$, are dominant. Furthermore, the power in this region is reactive.

Given these two regions, it seems obvious that antennas for RFID transponders should be designed to operate in the far-field region. However, the boundaries between far-field and near-field regions depend on the relationship between the physical dimensions of the antenna and the wavelength of the propagating signal. In particular, the dimensions of the antenna should be comparable to the wavelength of the signal in order to achieve optimal performance. The wavelength of signals operating in the LF region is around 2.4 km (1.5 miles), while the wavelength for signals operating in the HF region is 24 m (78 feet). Therefore, at these frequencies it would not be practical to build antennas with dimensions similar to the wavelength of their signals. Any antennas that can be built in a practical manner for RFID transponders operating in the LF or HF ranges will be electrically small and therefore highly inefficient. In contrast, RFID signals operating in the UHF region have a wavelength of approximately 30 cm. Although this dimension is still very large for building practical antennas, the following sections will explore how they can be modified to obtain antennas for UHF transponders that will operate in the far-field region.

RFID transponders operating in the LF or HF frequencies cannot use dipole antennas because of the mismatch in dimensions. An alternative solution to this problem is to use a small-loop antenna instead of a dipole. While the near fields radiated by a dipole are mainly electrical fields, the near field from a loop antenna is a magnetic field. A small-loop antenna is a closed loop with a maximum dimension that is less than about a tenth of the wavelength of the signal. It can be shown that the small-loop antenna is the dual equivalent of an ideal dipole, thus making it suitable for antennas incorporated in transponders operating at LF or HF frequencies.

Let's consider a conductor of infinite length carrying a magnitude of current of I amps. The magnetic field (B_ϕ) measured at a distance of r meters from the conductor can be found using Ampere's law as

$$B_\phi = \frac{\mu_0 I}{2 \pi r} \quad (Weber \,/\, m^2) \tag{2.1}$$

where μ_0 is the permeability of free space ($4 \pi\ 10^{-7}$ H/m).

However, a conductor of infinite length is not realistic. Its practical implementation is based on building a loop antenna by bending the original, finite wire, which carries a current of I amps in a circle with a radius of a meters. In practice, the wire is bent in such a way that produces a total of N turns as this allows using a longer wire with a relatively small diameter. In this situation, the value of the magnetic field in the z-coordinate direction (B_z) for a point located at a distance of r meters from the plane of the coil and located along the axis of the coil, as shown in Figure 2.1, can be found as

$$B_z = \frac{\mu_0 I N a^2}{2 (a^2 + r^2)^{3/2}} \quad (Weber \,/\, m^2) \tag{2.2}$$

where a is the radius of the loop in meters.

Assuming that the distance at which the field is measured, r, is much larger than the radius of the coil ($r^2 \gg a^2$), equation (2.2) can be simplified as follows:

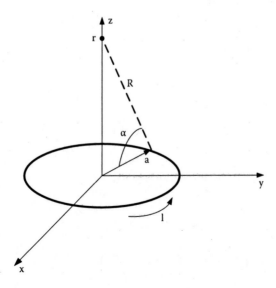

FIGURE 2.1
Magnetic field produced by loop antenna.

$$B_z = \frac{\mu_0 I N a^2}{2\, r^3} \quad (Weber\,/\,m^2) \tag{2.3}$$

As shown by equation (2.3), the magnetic field created by the small-loop antenna is dependent on $1/r^3$. This causes a strong decrease of field intensity as one moves away from the location of the coil antenna, and will seriously limit the read range of transponders using this type of antenna. However, even with this limitation, the small-coil antenna is more desirable than an electrically small dipole.

The antenna in the RFID transponder has to be both a receiving antenna and a transmitting antenna. It has to collect enough energy from the electromagnetic field generated by the interrogator to be able to power its integrated circuit, and it also has to transmit the data signal generated by the chip back to the interrogator. The *reciprocity theorem* indicates that the receiving and transmitting patterns for a given antenna are identical, and therefore the same antenna can be used for either task.

The voltage induced in a coil consisting of N loops immersed in a time-variant magnetic field can be calculated using Faraday's law as

$$V = -N\frac{d\Psi}{dt} \tag{2.4}$$

with Ψ being the magnetic flux.

The magnetic flux, in turn, can be calculated as

$$\Psi = \int B \bullet dS \tag{2.5}$$

where
  B = the magnitude of the time-variant magnetic field
  S = the surface area of the coil
  $\cdot$ = the inner product operator

Equation (2.5) indicates that the magnetic flux, and consequently the voltage induced in the coil, is dependent on the relative orientation between the coil and the magnetic field. This has some important implications at the time of using interrogators and transponders. Assuming that the magnetic field B was generated by the interrogator using another loop antenna, the maximum received voltage will occur when both coils are placed parallel to each other. Because the voltage generated in the coil that will be used to power the chip in the transponder is proportional to the intensity of the magnetic field (B), this voltage will also be dependent on $(1/r^3)$. This is important because it limits the maximum distance between the interrogator and transponder coils that will produce the threshold voltage necessary to power the chip on the transponder.

Although the previous equations describe the full underlying relationships between the variables involved in this process, it is more desirable from a practical point of view to study how these equations evolve in specific cases. Assuming that the coil is specifically tuned to the frequency of the time-variant magnetic field, the induced voltage in the coil is as follows:

$$V_0 = 2\pi f N S Q B_0 \cos\beta \tag{2.6}$$

where

V_0 = the induced voltage in the coil

f = the frequency of the time-variant magnetic field

B_o = the intensity of the field

N = the number of loops in the coil

Q = the quality factor of the coil antenna

S = the area of the coil

β = the angle between the plane of the coil and the direction of the field

Keeping in mind that the magnetic field generated by the coil in the interrogator is perpendicular to its plane, and that equation (2.6) will be maximized when $\beta = 0$, the induced voltage will be maximized when the interrogator and transponder coils are placed parallel to each other. Once again, the limiting factor in equation (2.6) comes from the intensity of the magnetic field (B_0) decreasing with the cube of the distance between coils ($1/r^3$).

2.1.2 Inductance of Coil Antennas Operating in the LF and HF Regions

Because, as shown in equation (2.6), the induced voltage is proportional to the quality factor (Q factor) of the coil antenna, it is desirable to increase the Q factor as much as possible. This can be achieved by carefully choosing the value of the inductance of the antenna. This section discusses the values of inductance found in typical antennas operating in the LF and HF regions.

Typical RFID systems use inductance values for antennas in transponders of few mH, while antennas for the interrogators have typical values between 10 and 100 times lower. It is necessary to keep in mind that the calculation of inductance values presented here can only be considered as approximated values. These calculations do not take into consideration the finite value of the conductivity of the wire that results in Ohmic losses, and, more importantly, they do not take into account the distributed stray capacitances that appear in the coil. A more detailed and comprehensive analysis and calculations of inductance values in antennas are beyond the scope of this book, but they can be found in any book on electromagnetism. If the designer requires an accurate value of inductance, the best option is to measure it with a complex impedance meter. This instrument is also able to measure the Q factor of the antenna, providing good characterization of the antenna.

The inductance of the coil shown in Figure 2.2, made with a single layer of wire and with n turns and a core of air, is approximately:

$$L = \frac{d^2\, n^2}{18d + 40l} \, (\mu H) \qquad (2.7)$$

where

    d = the diameter of the coil in inches

    n = the number of turns

    l = the length of the coil in inches

Equation (2.7) shows the square relationship between inductance and number of turns; if the number of turns is doubled, the resulting inductance is quadrupled.

Example 2.1:

Calculate the number of turns required to achieve an inductance of 10 μH by winding a piece of wire over an air core with a diameter of 1 inch. The length of the resulting coil must be equal to 1.25 inches.

Solution:

In this situation, l = 1.25 inches, and d = 1.0 inches. Therefore, we can write

$$10 = \frac{(1.0)^2\, n^2}{18 \cdot 1.0 + 40 \cdot 1.25}$$

Solving for n gives n = 26.07 turns. Given the limitations of the approximation for equation (2.7) plus the stray capacitance issues mentioned earlier, a good starting point would be to choose 26 turns. In this case, the value of inductance that we would measure in practice is likely to differ from 10 μH. The designer would have to adjust the number of turns after measuring the actual value of inductance. A recommended procedure is to start by choosing a larger number of turns and therefore a larger number of inductance, and proceed by removing turns until the coil achieved the required inductance.

Although this type of structure to create the antenna for an RFID system is not the most commonly used, it is nevertheless employed in some applications, especially in those that required a hermetically sealed transponder. Figure 2.3 shows an RFID transponder operating in the LF frequency in which it is possible to see the coil antenna and its windings.

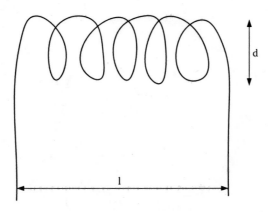

FIGURE 2.2
Single-layer, air-core coil.

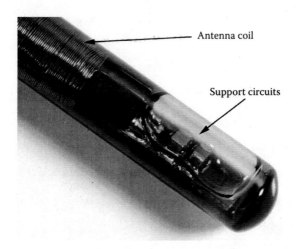

FIGURE 2.3
Detail of coil antenna in a commercial radiofrequency identification (RFID) transponder for the low-frequency (LF) range.

The practical problem associated with the use of this type of antenna is the three-dimensional space that it requires. Such antennas are not practical for applications that require a flat transponder, such as a credit card, and they are limited to applications that can support the tubular shape, for example in animal implantation. For this reason, the majority of transponders in the LF and HF frequency regions use antennas formed on a single plane, using several layers of conductor to achieve the desired inductance. Figure 2.4 shows this configuration for a multilayer circular coil with *n* turns.

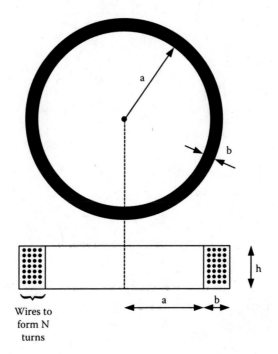

FIGURE 2.4
Multilayer, circular-coil, *N*-turn antenna.

The inductance of this coil is

$$L = \frac{0.31 a^2 n^2}{6a + 9h + 10b} \quad (\mu H) \tag{2.8}$$

where
   a = the radius of the coil in cm
   n = the number of turns
   b = the winding thickness of the coil in cm
   h = the winding height in cm

This configuration allows building coils with the required amount of inductance with minimal thickness, therefore minimizing the transponder's footprint. A variation of this antenna that results in an even thinner antenna is the spiral wound coil with a single layer shown in Figure 2.5.
The inductance of a single-layer spiral antenna is

$$L = \frac{0.3937 \, a^2 n^2}{8a + 11b} \quad (\mu H) \tag{2.9}$$

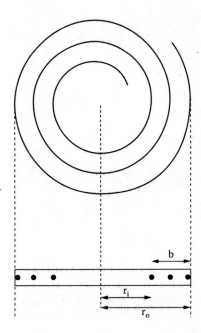

FIGURE 2.5
Single-layer, spiral-coil antenna.

where

a = the average radius of the coil in cm $(a = \frac{r_i + r_o}{2})$

b = ro − ri in cm

n = the number of turns in the spiral

This structure results in a thinner antenna, although the number of turns in the antenna is limited to the number of turns that is possible to pack in a single layer. This limits the inductance of the antenna as this value depends on the square of the number of turns. Other forms of planar coil antennas are the multilayer square-loop coil antenna with n turns and the multilayer rectangular-loop coil antenna, also with n turns, shown in Figure 2.6.

The inductance for the planar multilayer square loop coil antenna with n turns, shown in Figure 2.6(a), is

$$L = 0.008 \, a \, n^2 \left[2.3 \log\left(\frac{a}{h+b} \right) + 0.2235 \frac{b+h}{a} + 0.726 \right] (\mu H) \qquad (2.10)$$

where

n = the number of turns

a = the distance from the center of the coil to the side in cm

h = the thickness of the winding in cm

b = the width of the winding in cm

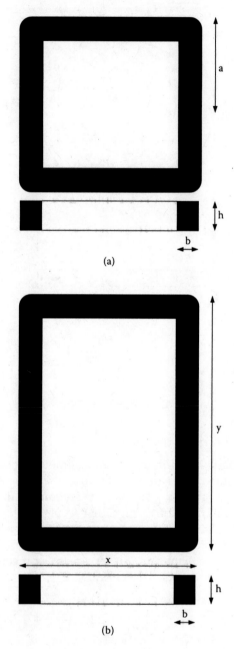

FIGURE 2.6
(a) Multilayer, square-coil antenna. (b) Multilayer, rectangular-coil antenna.

The inductance for the planar multilayer rectangular-loop antenna with n turns, shown in Figure 2.6(b), is

$$L = \frac{0.0276 \, n^2 \left(x + y + 2h\right)^2}{1.908 \left(x + y + 2h\right) + 9b + 10h} \quad (\mu H) \qquad (2.11)$$

where

x and y = the dimensions of the coil in cm

h = the thickness of the coil in cm

b = the width of the winding in cm

n = the number of turns

It is necessary to state once again that these equations, used to calculate the inductance of various coils are only approximations and need to be taken with some care. In particular, equation (2.11) is valid only when the dimensions x and y of the coil are quite different. For example, if the two dimensions of the coil in equation 2.11 (x and y) were identical, thus resulting in a square coil, the inductance values calculated by equations (2.10) and (2.11) will be different. These equations are provided as a starting point for the designer of antennas for RFID transponders.

Some antennas in transponders are configured as planar rectangular spiral structures, as shown in Figure 2.7. Figure 2.8 shows an example of a commercial transponder using this type of antenna.

There is not a full and closed equation to calculate the inductance of this type of antenna. However, its inductance can be calculated as the sum of the self-inductances for all the straight segments minus the sum of the mutual inductances existing in the arrangement. The mutual inductance is

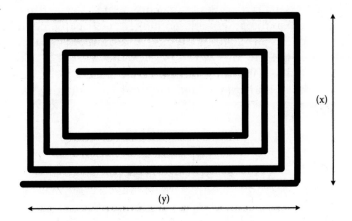

FIGURE 2.7
Single-layer, rectangular planar spiral antenna.

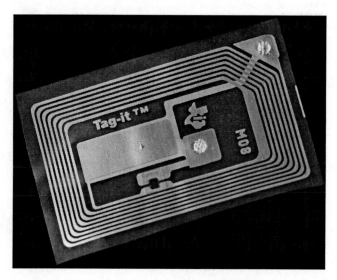

FIGURE 2.8
Commercial transponder using a planar, rectangular-spiral antenna.

the inductance that results from the magnetic fields produced by adjacent conductors. This mutual inductance will be positive when the direction of the current in the conductors is the same, and it will be negative when it is opposite.

2.1.3 Antennas for RFID Transponders Operating in the UHF Region

When the transponder is operating in the UHF region, the wavelength of the electromagnetic radiation is now comparable to the dimensions of the antenna. Because the antenna is no longer electrically small, a dipole can now be used. A *short dipole* is defined as a dipole whose length does not exceed one-fiftieth of the wavelength of the current signal. A short dipole, as depicted in Figure 2.9, produces an electric field that in near-field conditions has both radial and zenithal components, as given by equation (2.12):

$$E_r = \frac{2 I_0 l}{4\pi} k^2 \eta_0 \, e^{-jkr} \left(\frac{1}{(kr)^2} - \frac{j}{(kr)^3} \right) \cos\theta$$

$$\tag{2.12}$$

$$E_\theta = \frac{I_0 l}{4\pi} k^2 \eta_0 \, e^{-jkr} \left(\frac{j}{kr} - \frac{1}{(kr)^2} - \frac{j}{(kr)^3} \right) \sin\theta$$

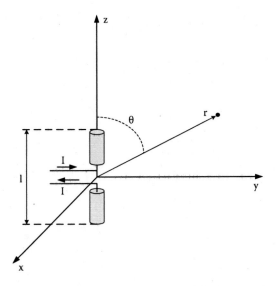

FIGURE 2.9
Electric field produced by a short dipole.

where
  k = the wavenumber $\left(k = \dfrac{2\pi}{\lambda} \right)$

  η_0 = the intrinsic impedance of free space $\left(\eta_0 = \sqrt{\dfrac{\mu_0}{\varepsilon_0}} \approx 120\pi \right)$

  r = the distance for the center of the dipole at which the field is measured

  θ = the zenith angle at which the field is measured

Assuming that the far-field approximation is valid ($r \gg \lambda$), the radial component of the electrical field can be neglected, and therefore equation (2.12) becomes

$$E_\theta = \frac{jI_0 l}{4\pi} k\eta_0 \frac{e^{-jkr}}{r} \sin\theta \qquad (2.13)$$

The short dipole, however, does not radiate effectively because it has a very small radiation resistance, normally below 1 Ω, making it unsuitable for a good impedance match. Therefore, other dipole alternatives are preferred when possible, such as the half-wave dipole. The *half-wave dipole*, depicted in Figure 2.10, receives its name because its length equals half of the wavelength of the current signal traveling through the antenna.

A half-wave dipole also produces an electric field only in the zenith component that can be described by equation (2.14):

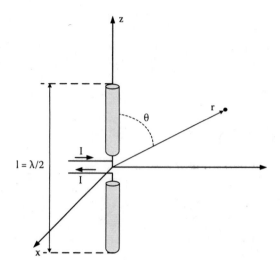

FIGURE 2.10
Electric field produced by a half-wave dipole.

$$E_\theta = \frac{j\,60 I_0}{r} \left(\frac{\cos\left(\frac{\pi}{2}\cos\theta \right)}{\sin\theta} \right) e^{-jkr} \tag{2.14}$$

One of the main advantages of the half-wave dipole is that its radiation resistance, around 73 Ω, is much larger than the radiation resistance of the short dipole. This makes the half-wave dipole better suited to match the impedance presented by its load, the integrated circuit in the transponder.

The wavelength for the typical UHF RFID frequency of 915 MHz is approximately 33 cm. Therefore, the length of a half-wave dipole is 16.5 cm. This value is valid only for free space; for an antenna mounted on an inlay, the actual value of its half wavelength decreases as the dielectric constant of the substrate increases. Although this distance is achievable, it presents some drawbacks. First of all, 16.5 cm may still be too large for some applications that require or would benefit from smaller transponders. More importantly, the power transfer characteristics for this antenna are very small and therefore require substantial matching approaches. Antenna matching is described in further detail in Section 2.2.

The space required by the antenna can be shortened while keeping the length of the conductors to the required 16.5 cm, for example by bending or even folding the wires that make up the dipole, as shown in Figure 2.11. This process can be repeated several times, bending the wires of the antenna in different directions as shown in Figure 2.12, which results in the structure known as the *meandered dipole*.

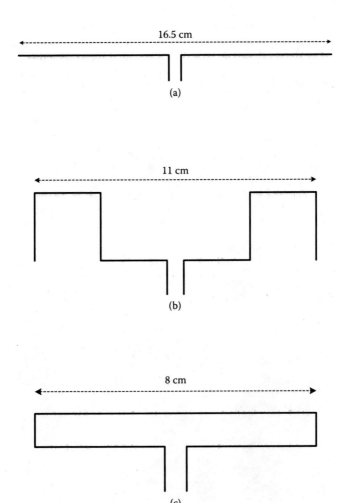

FIGURE 2.11
Strategies for reducing the size of ultra-high-frequency (UHF) antennas. (a) Original half-wave dipole. (b) Bent dipole. (c) Folded dipole.

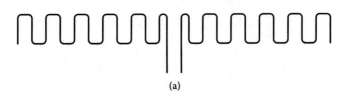

FIGURE 2.12
Examples of meandered dipole antennas. (a) Meandered dipole in a single direction.
(b) Meandered dipole in two directions.

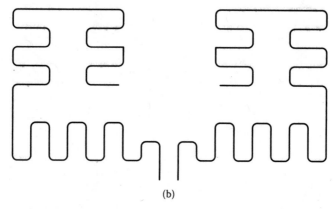

(b)

FIGURE 2.12 (Continued)
Examples of meandered dipole antennas. (a) Meandered dipole in a single direction.
(b) Meandered dipole in two directions.

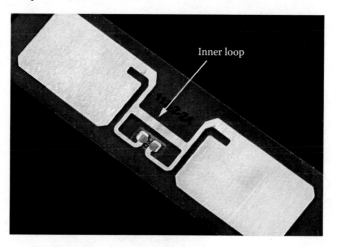

FIGURE 2.13
Detail of the inner loop used in some UHF antennas.

However, bending the dipole degrades the electrical characteristics of the antenna: the current lines flowing in opposite directions through adjacent paths have a canceling effect that in turn increases the radiation resistance of the antenna. Because this effect depends on the distance between the paths, the canceling effect becomes stronger when the antenna has multiple bends closely packed. In practice, the total length of a bent or meandered dipole is longer than the straight half-wave wavelength, although the total size of the antenna will be smaller.

Some antennas operating at the UHF range also incorporate a low DC resistance path closer to the center of the antenna known as the *inner loop*. This inner loop, shown in Figure 2.13, presents a series of advantages that makes its use common in this type of antenna.

First, the inner loop presents a low DC resistance that helps reduce the potential of damage to the integrated circuit due to high-voltage electrostatic discharges that may occur in the antenna or its vicinity. Second, the inductance of the inner loop can be used in conjunction with the input capacitance of the integrated circuit to form a resonance circuit that will couple with the magnetic near field, producing enough voltage to power the integrated circuit. Near-field communications for UHF RFID systems, although not very common yet, are growing rapidly. Third, the inner loop can be used like a transformer to help match the impedance of the integrated circuit to the antenna. Finally, the impedance of the antenna can be lowered or increased by changing the width of the trace in the inner loop.

2.2 Antenna Tuning for RFID Transponders

Antenna tuning for RFID transponders operating at LF or HF frequencies attempts to make the antenna resonant at the frequency of transmission, thus maximizing the current or the voltage in the system. Because the antenna has a marked inductive component, resonance is achieved by placing a capacitor of the adequate value in series or in parallel with the antenna.

When the capacitor is placed in series with the antenna, the resulting series resonant circuit has an impedance minimum at the resonance frequency and therefore maximizes the current being delivered to its load. This type of configuration is typical for interrogators and will be studied in Chapter 4. When the capacitor is placed in parallel with the antenna, the result is a parallel circuit that exhibits an impedance maximum at the resonant frequency and therefore maximizes the voltage across its terminals. This is the typical configuration for transponders. The tuning capacitor in RFID transponders can be created over the same substrate as shown in Figure 2.14.

The resonant circuit can therefore be modeled as a parallel circuit, similar to the one shown in Figure 2.15, in which L represents the inductance of the antenna, C represents the capacitance of the capacitor that must be added, R represents the resistance of the load that is the integrated circuit that must be powered, and the value r represents the resistance of the coil. Because normally $r \ll R$, the value r will be ignored in the calculations needed to find the value of required capacitance. However, the resistance r of the inductor contributes to degrading the overall performance of the antenna. These effects will be studied in the last section of this chapter.

The complex impedance presented by the circuit shown in Figure 2.15, neglecting the Ohmic losses r, is

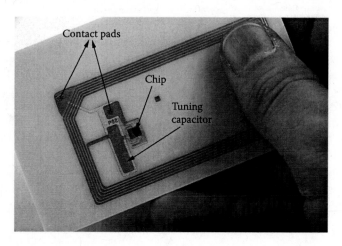

FIGURE 2.14
Tuning capacitor in the antenna of a high-frequency (HF) transponder.

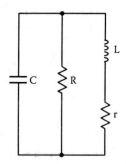

FIGURE 2.15
Parallel resonant model used to analyze antenna resonance. L: antenna inductance. C: external tuning capacitor. R: load (chip to energize). r: Ohmic losses in the antenna.

$$Z(j\omega) = \frac{\dfrac{j\omega}{C}}{\left(\dfrac{1}{LC} - \omega^2\right) + j\dfrac{\omega}{RC}} \tag{2.15}$$

where $\omega = 2\pi f$.

The resonant frequency for the parallel circuit is the frequency that creates the maximum for $Z(j\omega)$. This frequency is

$$\omega_o = \sqrt{\frac{1}{LC}} \quad or$$

$$f_o = \frac{1}{2\pi\sqrt{LC}} \tag{2.16}$$

From equation (2.15), it can be seen that at the resonant frequency, the impedance of the parallel circuit is real and equal to the value of the resistance of the load. This is the maximum value for the impedance.

Example 2.2:

Consider the value of inductance for the antenna in Example 2.1 (10 μH). Calculate the value of parallel capacitance required to make the antenna resonant at the typical RFID frequencies of (a) 125 kHz for LF systems and (b) 13.56 MHz for HF systems.

Solution:

Solving for C in equation (2.16) yields

$$C = \frac{1}{L(2\pi f)^2}$$

Therefore, the values of capacitance required for resonance are as follows:
$C_{125\,kHz} = 162\ nF$
and
$C_{13.56\,MHz} = 13.7\ pF$

The same calculations for resonant frequency can be applied to the devices that communicate data by tuning and detuning the antenna. This can be accomplished by using, for example, two capacitors or two inductors in the parallel RCL circuit. In order to detune the antenna, the controlling device can, for example, shorten one of the capacitors or inductors, resulting in a new resonant frequency. This, in turn, results in a minimal transfer of energy back to the interrogator.

The devices MCRF355 and MCRF360 from Microchip Technologies operate following this principle. Figure 2.16 shows one of the configurations used. Here, the antenna will be tuned when both inductors are active and will be detuned when the device shortens the inductor L_2.

Using equation (2.13), the resonant frequency when the antenna is tuned is $f_{tuned} = 1/2\pi \sqrt{L_T\,C}$, with L_T being the total inductance between point A and ground. When the device detunes the antenna by shortening inductor L_2, the new resonant frequency becomes $f_{detuned} = 1/2\pi \sqrt{L_1\,C}$. Because $L_T > L_1$ ($L_T = L_1 + L_2 + 2k\sqrt{L_1 L_2}$, with k being the coupling coefficient between the coils), $f_{tuned} < f_{detuned}$.

The difference between the tuned and detuned frequencies must be chosen to maximize the Modulation Index and the read range. If it is necessary to choose the tuned and detuned frequencies such as $f_{tuned} < f_{detuned}$, this can be achieved by using the configuration shown in Figure 2.17.

The operation is similar as shown before. Here, $f_{tuned} = 1/2\pi \sqrt{L\,C_T}$, with $C_T = C_1 C_2 / C_1 + C_2$. The device detunes the circuit by shortening the capacitor

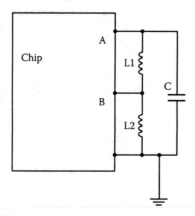

FIGURE 2.16
Tuned and detuned circuit using two inductors and one capacitor. This configuration results in $f_{tuned} < f_{detuned}$.

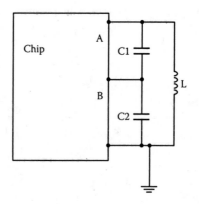

FIGURE 2.17
Tuned and detuned circuit using two capacitors and one inductor. This configuration results in $f_{tuned} > f_{detuned}$.

C_2, resulting in the following detuned frequency: $f_{detuned} = 1/2\pi \sqrt{L C_1}$. Because $C_T < C_1$, this results in $f_{tuned} < f_{detuned}$.

Equation (2.15) can also be used to calculate the bandwidth of the system defined as its −3 dB cutoff. This is

$$B = \frac{1}{2\pi RC} \quad (Hz) \tag{2.17}$$

An important parameter in a resonant circuit is its quality factor (Q), which compares the energy stored in the circuit with the energy that it dissipates:

$$Q = \frac{f_o}{B} \qquad (2.18)$$

Equation (2.17) can be rewritten taking into account the values of components in the resonant circuit by combining equations (2.16) and (2.17):

$$Q = R\sqrt{\frac{C}{L}} \qquad (2.19)$$

Equation (2.6), found earlier, was used to calculate the voltage induced in the coil. Substituting equation (2.19) into equation (2.6) yields

$$V_0 = 2\pi f N S Q B_0 \cos\beta \; = \; 2\pi f N S R \sqrt{\frac{C}{L}} \, B_0 \cos\beta \qquad (2.20)$$

This equation indicates that the voltage induced in the antenna transponder is inversely proportional to the square root of its inductance, and proportional to both the number of turns in the coil and its surface.

2.3 Antenna Matching for RFID Transponders

The purpose of matching an antenna to its load is to ensure that the antenna transfers the maximum amount of power to its load. In an RFID transponder, the load of the antenna is the integrated circuit that will be powered by the energy supplied by the antenna. Antenna matching is generally based on altering the complex impedance presented by the antenna at its operating frequency, by modifying its physical dimensions, inserting a reactive component, or a combination of both.

Consider a simple circuit, as shown in Figure 2.18. The power transferred by the voltage source V_s to the load resistance R_L is

$$P_L = \frac{R_L}{(R_L + R_s)^2} \, V_s^2 \qquad (2.21)$$

The goal of power matching is to find a relationship between R_s and R_L that will maximize the power transferred to the load. In mathematical terms, this

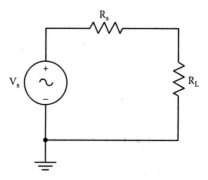

FIGURE 2.18
Simplified model for power transfer calculations.

is equivalent to finding the relationship between R_L and the rest of the circuit
elements that will make the first derivative of equation (2.21) zero. That is,

$$\frac{dP_L}{dR_L} = \frac{(R_L + R_s)^2 - 2R_L\,(R_L + R_s)}{(R_L + R_s)^4}\,V_s^2 = 0 \quad \Rightarrow \quad R_L = R_s \qquad (2.22)$$

In a transponder, the voltage source models the voltage induced in the
antenna, and the load models the integrated circuit that must be powered.
These are not purely resistive elements such as those depicted in Figure 2.18,
but they have resistance and reactance. Therefore, a more accurate model for
this situation is shown in Figure 2.19.

Equation (2.23) shows the power transferred to the load that is the real part
of the impedance of the integrated circuit:

$$P_L = \frac{R_{ic}}{2\left[\left(R_{ant} + R_{ic}\right)^2 + \left(X_{ant} + X_{ic}\right)^2\right]}\,V_{ant}^2 \qquad (2.23)$$

There are now two conditions that must be met in order to transfer the
maximum amount of power between the source (voltage generated by the
antenna) and the load (integrated circuit to be powered). These are

$$\begin{cases} R_{ic} = R_{ant} \\ X_{ic} = -X_{ant} \end{cases} \qquad (2.24)$$

Equation (2.24) can now be understood by saying that the amount of power
transferred from the antenna to the integrated circuit will be maximum when
their impedances are conjugate. This is an advantageous condition for RFID
systems as the impedance of the antenna has an inductive behavior and the

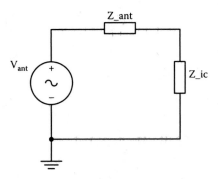

FIGURE 2.19
Model for power transfer calculations in an RFID transponder.

input impedance of the integrated circuit presents a capacitive behavior. The model shown in Figure 2.19 can be further refined as the model shown in Figure 2.20.

It is necessary to point out that the series impedance model used in Figure 2.20 is not the most common way to represent the input impedance of an integrated circuit or any other electronic device. Generally, input impedance is expressed using a parallel model with the input resistance and the input capacitance. However, to preserve the parallelism used in Figures 2.18 and 2.19, it is more convenient to express it using the series model. Given the values of parallel resistance and capacitance, it is possible to find their equivalent series values using equation (2.25):

$$\begin{cases} R_s = \dfrac{R_p}{1+\left(\omega R_p C_p\right)^2} \\[4mm] C_s = \dfrac{1+\left(\omega R_p C_p\right)^2}{\omega^2 R_p^2 \, C_p} \end{cases} \qquad (2.25)$$

Example 2.3:

The specifications for the chip MCRF200 from Microchip used in RFID transponders indicate a parallel capacitance of 2 pF. Assuming a parallel resistance of 3 kΩ, find the equivalent series model at the frequency of 915 MHz.

Solution:

Equation (2.25) gives the following values of the equivalent series model:
R_s = 2.6 Ω and C_s = 1.96 pF. In general, the values of series and parallel capacitance will be approximately the same. However, the values of the series and parallel resistances are very different for each model.

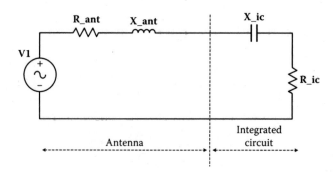

FIGURE 2.20
Series model for the transponder antenna and chip to be energized.

It is also interesting to calculate the gain or attenuation between the voltage delivered to the integrated circuit and the voltage generated by the antenna in the optimal case of maximum power transfer. Using the model shown in Figure 2.16, the voltage across the integrated circuit produced by the antenna can be calculated as follows:

$$V_{ic} \approx j \frac{1}{2\pi\omega\, R_{ic}\, C_{ic}} V_{ant}$$
(2.26)

Example 2.4:

Using the values given in Example 2.3, calculate the voltage gain or attenuation at the input of the integrated circuit at the frequency of 915 MHz.

Solution:

Using equation (2.26), V_{ic} / V_{ant} is equal to 34. This means that the voltage at the input of the integrated circuit in the transponder at the resonance frequency will be 34 times larger than the voltage generated by the antenna. In practice, this number will be lower because equation (2.26) has not considered factors such as the effect of the Q factor and other losses in the system. Typical voltage gains in practice are around 10 to 15. In any case, this represents an important benefit for the designer of the RFID system.

It is necessary to consider that, in practice, there will be some mismatch between the impedance of the integrated circuit and the impedance of the antenna as these will not be perfect conjugates. This mismatch will produce losses in the power being delivered to the integrated circuit, resulting in decreasing the sensitivity and read range for the transponder. This factor becomes more important for UHF antennas designed for global operations that have a frequency bandwidth of 860 MHz to 960 MHz rather than those designed for operation at a single frequency. In this case, the usual design approach is to place the matched frequency at the geometrical mean of the two frequency ends, resulting in 908 MHz. This will result, however, in mismatch losses at frequencies close to 860 MHz and frequencies close to 960 MHz.

2.4 Antennas in Commercial RFID Transponders

Coil antennas are used in the different type of encapsulated glass transponders, similar to Figure 2.21, operating in the LF region manufactured by Texas Instruments. Glass transponders are hermetically sealed, making them waterproof and therefore suitable to be immersed in water and other fluids.

These transponders are marketed toward access control, vehicle identification, container tracking, asset management, animal identification, and waste management applications, as described by the literature provided by their manufacturer. Texas Instruments has different versions of these transponders depending on the size of the memory in the chip and its ability to be rewritten by the user. The transponders are also available in different sizes, in particular lengths of 12 mm, 23 mm, and 32 mm. Figure 2.22 shows the different sizes of these glass transponders that operate at the central frequency of 134.2 kHz, corresponding to the LF range. Other operational characteristics of these glass transponders are listed in Table 2.1.

FIGURE 2.21
Glass transponder used for RFID applications in the LF range.

FIGURE 2.22

Three commercial glass transponders of different sizes used for application in the LF range.

TABLE 2.1

Operational Characteristics of Glass Transponders

Main Operational Parameters Glass Transponders Texas Instruments – LF Region	
Modulation	FSK: 134.2 kHz and 123.2 kHz
Transmission Principle	Half Duplex
Read Range	Less than 100 cm
Maximum Operating Temperature Range	− 25°C to + 85°C
Read Time	Less than 70 ms
Case material	Glass
Protection	Hermetic seal
Weight	0.8 grams
Memory on board	4 different options depending on specific part number

Other transponders operating in the LF region use planar coil antennas similar to those shown in Figure 2.23. The dimensions of the larger transponder are 25 mm for the outer ring and 20 mm for the inner ring, while the dimensions for the smaller antenna are 14 mm for the outer ring and 9 mm for the inner ring.

These antennas consist of a length of conductor wire coiled multiple times in order to create the desired inductance. The manufacturer specifies their use for the same general type of applications as the glass transponders, with the advantage of being much thinner.

Antennas for transponders operating in the HF frequency (13.56 MHz) are normally planar antennas, having a structure similar to the planar antennas used in the LF region. However, because HF antennas require a lower

FIGURE 2.23
Commercial planar coil antennas.

inductance, the number of turns for these antennas is much lower, normally less than 10 turns. Planar antennas for RFID transponders operating in the HF region are also manufactured in different sizes and thicknesses depending on the intended application. Figure 2.8 showed a picture of an HF transponder manufactured by Texas Instruments under the name of Tag-it™. This type of transponder is marketed toward product authentication, library applications, supply chain management, asset management, and ticketing or stored value applications. The antennas are built over a polyethylene-terephthalate (PET) substrate giving a maximum thickness of 0.085 mm for the antenna area and 0.355 mm for the chip area. Figure 2.24 shows three antennas from this same family. The larger rectangular antenna has maximum dimensions of 75 mm × 45 mm and consists of seven turns; the small rectangular antenna has a maximum dimension of 39 mm × 22 mm and has six turns; the square antenna has maximum dimensions of 45 mm × 45 mm and contains nine turns.

The number and type of different antennas for transponders operating in the UHF region are higher as each commercial product has been optimized with an application in mind in terms of the different antenna parameters or the type of substrate better suited for a specific purpose. Figures 2.25, 2.26, and 2.27 show three examples of the diverse UHF transponders manufactured by Alien Technology.

Figure 2.25 shows the Alien ALN-9640 Squiggle® Inlay from Alien Technology. This is a general-purpose transponder used for packages that

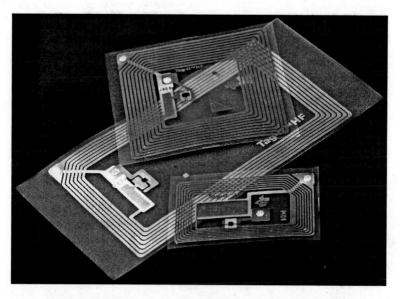

FIGURE 2.24
Diverse planar rectangular antennas used in commercial HF transponders.

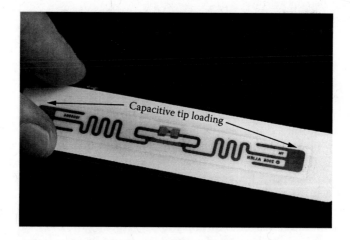

FIGURE 2.25
Commercial UHF transponder based on a meandered dipole and using capacitive tip loading.

include metal or water. The largest length dimension of this antenna is 98 mm, and the largest width of the overall antenna is 8 mm. This antenna exhibits a radiation pattern similar to a half-wave dipole. This antenna is based on a half-wave dipole that has been made shorter by bending the wires and adding shunt and series inductors to create the adequate resonant frequency for the integrated circuit that will be powered by the antenna. The large conductive areas located at each end of the antenna serve as capacitors used to increase the antenna capacitance. This technique is known as *capacitive tip loading*. Since the magnitude of capacitive reactance decreases as the capacitance increases, a tip-loaded dipole exhibits a more inductive behavior than a dipole of the same length without the additional capacitance at its ends.

Figure 2.26 shows the Alien ALN-9534 2x2 Inlay from Alien Technology. This antenna exhibits a more omnidirectional radiation pattern and is marketed toward item-level tagging applications such as apparel or baggage. The maximum dimensions of this antenna are 46 mm × 40 mm. This type of antenna, in which the conductors are thicker than the previously studied wire antennas, presents the advantage of having higher capacitance and lower inductance than the wire antennas, therefore making it easier to match to the characteristics of its integrated circuits. Furthermore, these antennas made with thicker conductors exhibit a wider bandwidth than the wire antennas.

Figure 2.27 shows the Alien ALN-9554 M Inlay from Alien Technology. This antenna exhibits a better omnidirectional radiation pattern and is marketed toward tagging plastic totes, pallets, and other reusable assets. The maximum dimensions of this antenna are 98 mm × 38 mm. This type of antenna presents the higher bandwidth for the same reasons described earlier. When using conductive inks to create the antenna, the additional ink required to create them results in a cost increase for this type of antenna.

Figure 2.28 shows an example of a meandered dipole antenna with an inner loop for use in the UHF range manufactured by RSIID technologies.

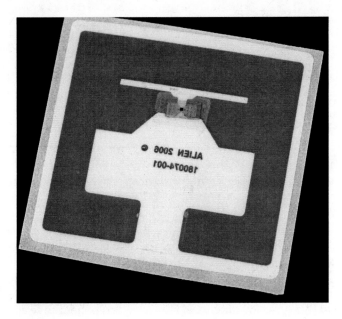

FIGURE 2.26
Commercial UHF transponder using a more omnidirectional antenna.

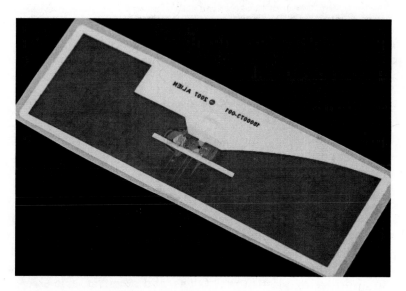

FIGURE 2.27
Commercial UHF transponder having a higher bandwidth response.

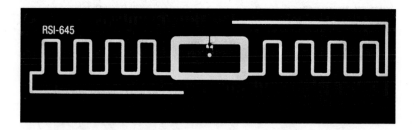

FIGURE 2.28
Meandered dipole antenna for UHF applications.

The maximum dimensions of this antenna are 67 mm × 13.5 mm, thus making it suitable for applications with limited available space. Figure 2.28 is just one example of the different configurations available for meandered dipoles.

Figure 2.29 shows a dual dipole transponder. These transponders have two dipole antennas to make them more insensitive to the orientation of the transponder in reference to the interrogator. The potential drawback of this approach is the additional space required by the transponder that may not be suitable for all products. Furthermore, the front end of the integrated circuit that will be powered by the antenna increases in complexity as it needs to respond to two different antennas. However, for longer range applications in which the position of the object can change, this type of antenna presents a clear advantage.

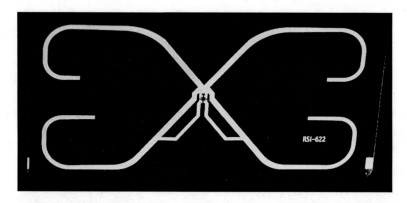

FIGURE 2.29
Dual dipole antenna for UHF transponders.

While the majority of transponders operating in the UHF range are powered by far-field electromagnetic radiation, in some cases the relatively large area that they require may be larger than the available surface in the transponder. The solution in this case is to use UHF antennas that operate in the near-field region, accepting the drawbacks associated with this mode of operation. These antennas operate based on the same inductive coupling principles used by LF and HF antennas. However, it is possible to add shunt and series inductors or capacitive tip loads to create what are known as *hybrid antennas*. These hybrid antennas, examples of which are shown in Figures 2.30 and 2.31, are able to operate in the near and far fields even though the performance in the far-field region is degraded compared to the performance of a typical dipole antenna.

2.5 The Connection between the Chip and the Antenna in RFID Transponders

A critical step in the manufacture of RFID transponders is the connection between the antenna and the integrated circuit in the transponder. There are main methods of doing this connection: chip-on-board and direct die attachment.

Chip-on-board (COB) is an assembly technology for semiconductors in which the electronic chip is directly mounted on and connected to its final board instead of undergoing the traditional packaging process for integrated circuits. This process reduces space requirements and cost, and also increases the performance of the system due to the decrease in connection lengths and increase in reliability. This approach is normally used for wire-wound antennas. During the COB process, the capacitor used to create resonance is also packaged in the device. The device is then covered in epoxy in such a

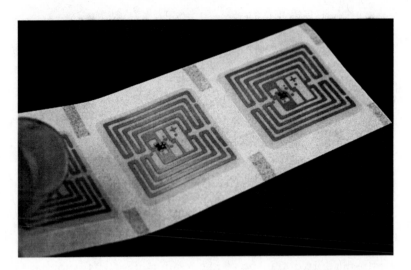

FIGURE 2.30
Hybrid UHF antenna able to communicate in near field and far field.

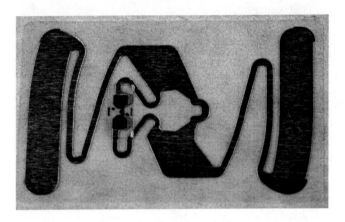

FIGURE 2.31
Hybrid UHF antenna with tip loading.

way that only the two cables to connect to the antenna emerge from the package, as shown in Figure 2.32. The majority of COB packages are used in cards that need to meet the requirements for standard thickness of these cards at 0.76 mm. With this requirement, the typical thickness of COB packages is about 0.41 mm. Although the COB package is designed to protect the internal silicon device during the card lamination process, it is necessary to be careful to prevent mechanical cracks on the device as a result of the mechanical pressures and hot temperatures.

Direct die attachment can be achieved using two techniques: wire bonding or flip-chip. Flip-chip is a technology for interconnecting semiconductor

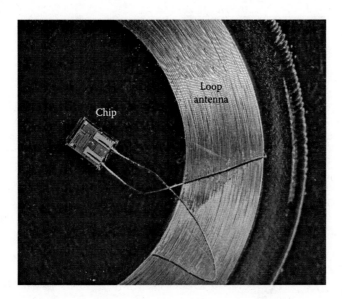

FIGURE 2.32
Connection of the antenna wires to the chip in the transponder.

devices by creating conductive bumps in the pads of the chip. In order to connect the chip, this has to be flipped so the bond pads are in contact with the matching pads in the circuit. In the wire-bonding technology, the chip is mounted upright and wires are used to connect the chip to the circuit. In the case of RFID transponders, the bumps or wires are connected to the antenna, thus being an ideal method to use for printed, etched, or stamped antennas. In this case, the resonance capacitor can also be etched on the substrate of the transponder, and therefore this technology does not require binding it with the chip as COB does. Figure 2.33 shows an example of this type of connection.

The choice between printing, etching, or stamping the antenna is a trade-off between cost and performance. The Q factor of the antenna, which as discussed earlier has a strong effect on the read range for the transponder, is inversely proportional to the resistance of the antenna. With this in mind, it is desirable to use an etched antenna instead of a printed antenna made of conductive material. The etching process is also less expensive than the printing process. For large antennas, however, etching and stamping waste too much unwanted material. In this case, a printed antenna becomes a more desirable solution. Figure 2.34 shows the general structure of the chip-to-antenna connection, independently of the specific technique being used. The antenna is connected between one of the terminals in the die and its ground connection (GND).

This connection creates a problem due to the crossing of the connection to the antenna with the ground ring around the die, creating a parasitic

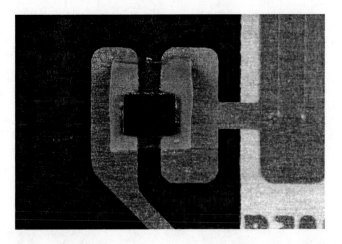

FIGURE 2.33
Direct connection between the chip and the antenna in the transponder.

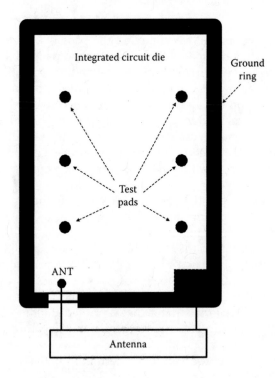

FIGURE 2.34
Strategy for reducing the stray capacitance when connecting the chip to the antenna in an RFID transponder.

capacitance. Depending on the technology being used to make this connection, the parasitic capacitance can have a value of up to 20 fF, influencing the resonant frequency for the antenna. To limit this effect, the ground ring around the die is narrowed in the vicinity of the pad for the antenna connection, thus reducing the parasitic capacitance.

2.6 Additional Factors That Affect the Performance of Antennas in RFID Transponders

As discussed in the preceding sections, the quality factor of an antenna is influenced by its Ohmic losses. For the coil antennas used for transponders operating in the LF region, these Ohmic losses come from the wire used to build the coil. The resistance in DC of a conductor of length l, with a uniform cross-section S, is

$$R_{DC} = \frac{l}{\sigma S} \qquad (2.27)$$

where σ is the conductivity of the material.

Because of the need to manufacture the antenna as small as possible, designers opt for choosing the narrowest possible diameter, resulting in an increase in its resistance as shown by equation (2.27). More important, however, for the design of antennas is their losses in AC. The density of AC current through a wire is not constant through its surface, but the current tends to flow through the outer section of the wire, thus decreasing its effective area. This is known as the *skin effect*. Similarly, the *skin depth* for a conductor is defined as the value of its depth at which the current density falls to 37% of the current density along the surface. As shown in equation (2.28), for a given material, its skin depth is dependent on the frequency of the AC current:

$$\delta = \frac{1}{\sqrt{\pi \mu \sigma f}} \qquad (2.28)$$

where
  δ = the skin depth
  μ = the permeability of the material
  σ = the conductivity of the material
  f = the frequency of the AC current through the material

Example 2.5:

Calculate the skin depth for copper wire at (a) 125 kHz and (b) at 13.56 MHz.

Solution:

Copper has a relative permeability of 1, so its total permeability is $\mu = 4\pi\ 10^{-7}$ H/m. Its conductivity is equal to $\sigma = 5.8\ 10^{7}$ mho/m. Therefore, equation (2.28) can be rewritten for copper as

$$\delta = \frac{0.066}{\sqrt{f}} \quad (m)$$

A. The skin depth of copper at 125 kHz is equal to 0.186 mm. This means that the current density of a point located 0.186 mm away from the surface of the conductor is only 37% of the current density on the surface.
B. The skin depth of copper at 13.56 MHz is equal to 0.018 mm. At this frequency, the skin depth is about one-tenth of the skin depth at 125 kHz. Therefore, at 135 MHz the majority of the current is flowing through the surface of the conductor.

The total resistance of a wire to AC current (R_{ac}) is a combination of its DC resistance (R_{DC}) and its skin depth (δ):

$$R_{ac} = R_{DC}\frac{a}{2\delta} \qquad (2.29)$$

where a is the radius of the wire.

For etched coil antennas, commonly found in planar transponders, the AC resistance is given by equation (2.30):

$$R_{ac} = \frac{1}{(w+t)}\sqrt{\frac{\pi\mu f}{\sigma}} \qquad (2.30)$$

where
  w = the width
  t = the thickness of the conductor on the substrate

Etching is a wasteful process as metal is removed from the substrate in order to create the antenna. Although some of the materials may be recycled, etching also requires large amounts of chemicals and energy to use in the process. To respond to this concern, newer processes based on conductive inks or copper deposits have been developed, especially for antennas operating in the UHF frequency range. Because of the higher conductivity of copper compared to silver-based conductive inks, antennas made with deposited copper can be substantially thinner than those made with conductive inks. In general, both materials can be effectively used to create antennas.

3

Transponders

CONTENTS

The two main components of a transponder are its internal circuits and its antenna. The antenna, studied in Chapter 2, is used to collect energy from the electromagnetic fields in which the transponder is embedded as well as to transmit the information back to the interrogator. The integrated circuit, also known as the *chip* or *device*, is the key component in the transponder, and for this reason this chapter uses the names *transponder*, *chip*, and *device* interchangeably. The integrated circuits have the ability to store information to be transmitted to the interrogator, execute a series of commands, and, in some cases, store new information sent by a remote station.

This chapter describes the fundamental blocks that make up the transponder. Because there are a large number of transponders in the market, each one with its own specific differences, this chapter focuses on the functional blocks common to them using commercial transponders to illustrate these concepts. Therefore, the devices used in the examples should not be seen as an exhaustive list of transponders but just as examples used to illustrate specific concepts and ideas. This chapter starts by describing the analog front end commonly encountered in these devices. The analog front end contains the radiofrequency section that is necessary to match the electrical properties of the device to the electrical properties of the antenna as well as the circuits used for harvesting and managing the power necessary to turn on the device. This section is followed by the description of the main encoding and modulation methods commonly used by transponders. This is followed

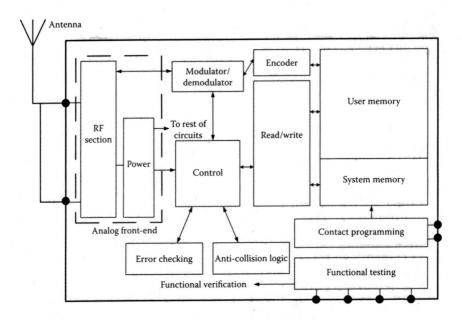

FIGURE 3.1
Block diagram of a generic RFID transponder highlighting its main functional components.

by the description of the different memory structures to store data and configuration parameters as well as the codes used to ensure the integrity of the digital data. The next section briefly describes the methods for programming the transponder by either a contact programmer or a contactless programmer. The last two sections in this chapter present a summary of the electrical and mechanical specifications found in commercial transponders as well as a description of their most common assembly methods.

Figure 3.1 depicts the block diagram of a basic and generic RFID transponder showing its main functional blocks. It is important to note that not all the blocks shown in the figure will be always present in any transponder depending on its use, intended application, and cost.

3.1 The Analog Front End

3.1.1 Radiofrequency Stage

When the transponder is immersed in an electromagnetic field of the appropriate frequency, a radiofrequency voltage appears across the antenna terminals. The task of the front-end stage is to rectify that radiofrequency voltage and convert it into a continuous voltage (DC) with a value high enough to power the rest of the circuits inside the device. Different transponders have different requirements regarding the minimum voltage required to become operational depending on their internal structure and fabrication technology.

The first step in the energy conversion process between the electromagnetic field and the DC voltage is the resonant circuit tuned to the frequency of the field. The different possible configurations for the external resonant circuit described in Section 2.2 are now studied, taking into account the structures used in commercial transponders. Figure 3.2 shows the possible alternatives for the external resonant circuit using two commercial devices: the MCRF355 and the MCRF450, both from Microchip Technology and both operating at 13.56 MHz.

Some devices incorporate one or two capacitors inside their integrated circuit, eliminating the need for using an external capacitor and therefore reducing the size required for the overall transponder inlay. Figure 3.3 shows the resonant circuit using commercial devices that incorporate one or two capacitors in their internal circuitry.

The devices MCRF360, MCRF451, and MCRF455 from Microchip Technology have a single internal capacitor with values of 100 pF, 95 pF, and 50 pF, respectively. The device MCRF422, also from Microchip Technology, has two internal capacitors of 50.6 pF and 65.4 pF. All these devices operate at 13.56 MHz.

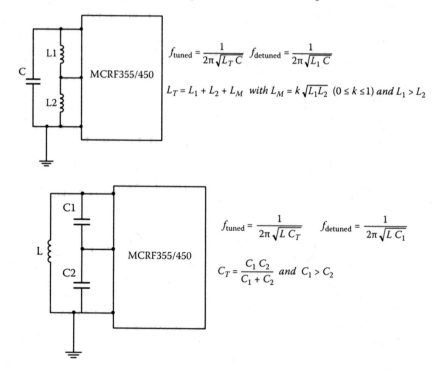

$$f_{\text{tuned}} = \frac{1}{2\pi \sqrt{L_T\, C}} \qquad f_{\text{detuned}} = \frac{1}{2\pi \sqrt{L_1\, C}}$$

$$L_T = L_1 + L_2 + L_M \quad \text{with } L_M = k\sqrt{L_1 L_2} \ \ (0 \le k \le 1) \ and \ L_1 > L_2$$

$$f_{\text{tuned}} = \frac{1}{2\pi \sqrt{L\, C_T}} \qquad f_{\text{detuned}} = \frac{1}{2\pi \sqrt{L\, C_1}}$$

$$C_T = \frac{C_1\, C_2}{C_1 + C_2} \quad and \ C_1 > C_2$$

FIGURE 3.2
Alternatives for the external resonant circuit found in commercial transponders. These transponders require a capacitor in the inlay to create the resonant circuit. The inductor in the circuits models the antenna.

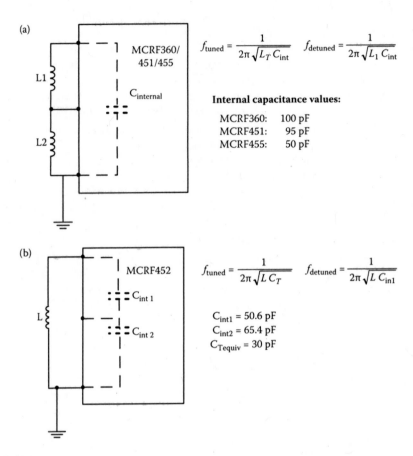

$$f_{tuned} = \frac{1}{2\pi\sqrt{L_T C_{int}}} \qquad f_{detuned} = \frac{1}{2\pi\sqrt{L_1 C_{int}}}$$

Internal capacitance values:

MCRF360:	100 pF
MCRF451:	95 pF
MCRF455:	50 pF

$$f_{tuned} = \frac{1}{2\pi\sqrt{L C_T}} \qquad f_{detuned} = \frac{1}{2\pi\sqrt{L C_{in1}}}$$

$C_{int1} = 50.6$ pF
$C_{int2} = 65.4$ pF
$C_{Tequiv} = 30$ pF

FIGURE 3.3
Alternatives for the external resonant circuit for devices that incorporate the resonance capacitor in the die of the integrated circuit. (a) Structure with a single internal capacitor. (b) Structure with two internal capacitors.

The path between one of the antenna connection nodes and ground, shown in Figures 3.2 and 3.3, is shortened according to a control signal. The path is shortened using a transistor, called a *modulation transistor*, that exhibits a low ON resistance, normally below 5 Ω and a very high OFF resistance of several MΩ. In this situation, the transponder exhibits two different resonant frequencies.

The resulting resonance frequency when the modulation transistor is OFF is chosen to be equal to the frequency of the electromagnetic field emitted by the interrogator. As shown in Figure 3.4, because of the resonance condition, the energy of the field is transmitted through the front end to the device, resulting in the radiofrequency voltage at the input of the device being maximal. This situation is called *uncloaking*. When the modulation transistor is turned ON, it shortens one of the inductors or capacitors, resulting in a resonant frequency for the circuit different from the frequency of the electromagnetic field. When it shortens one of the two inductors connected

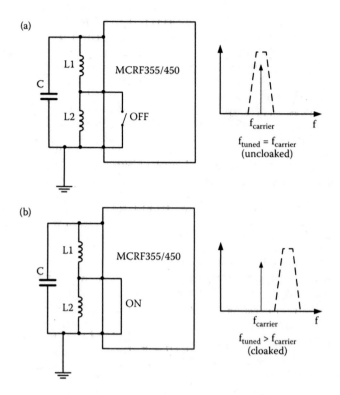

FIGURE 3.4
Cloaking and uncloaking modes produced by the control transistor in the transponder and the different resonant frequencies generated. (a) Uncloaked. (b) Cloaked.

in series, the resulting resonant frequency is higher than the frequency of the field; when it shortens one of the two capacitors connected in series, the resulting resonant frequency is lower. In any case, the frequency of the circuit and the frequency of the field are different, and therefore most of the energy of the electromagnetic field is rejected by the filter, resulting in the voltage generated at the input of the device being minimal, ideally zero. This situation is called *cloaking*. The same cloaking–uncloaking approach is used by the transponder to transmit data to the interrogator. Using this approach, the digital signal controls the modulation transistor.

Example 3.1:

Calculate the values of the external inductors needed to use in a transponder based on the MCRF451 device.

Solution:

The MCRF451 has an internal capacitance of 95 pF and operates at a frequency of 13.56 MHz. Using the equations shown in Figure 3.3, the total inductance necessary is $L_T = 1.45$ µH. With this inductance, the circuit resonates at 13.56 MHz, and therefore the energy from the electromagnetic field is passed to the device. The $L_T = 1.45$ µH of external inductance will be split between L_1 and L_2.

The value of L_1 can be calculated using the equations from Figure 3.3 now for the case of detuned frequency. The detuned frequency should be between 3 MHz and 6 MHz away from the tuned frequency. For the structure used by the MCRF451 device, the detuned frequency will be higher than the tuned frequency. Using for example a frequency shift of 4 MHz, the detuned frequency becomes 17.56 MHz, resulting in a value of L_1 equal to 864 nH. Assuming initially a value for the coupling coefficient between L_1 and L_2 equal to zero ($k = 0$), the required value for L_2 is 586 nH. It can be observed that $L_1 > L_2$ as it is required. In a more realistic situation, the value of the coupling coefficient will be between 0 and 1, resulting in a lower value for L_2.

3.1.2 Power Management

In order to be operative, the transponder needs to convert the radiofrequency voltage detected by the antenna into a DC voltage. The voltage required to bias the internal components in the transponder is higher than the voltage detected by the antenna. Therefore, the transponder requires the use of voltage multipliers to reach the values necessary by the biasing voltage.

A voltage multiplier is a circuit that converts a lower AC voltage into a higher DC voltage. Figure 3.5 shows the basic structure of a simple voltage multiplier called a *voltage doubler*.

Assuming initially ideal diodes for simplicity, diode D1 is forward biased during the phase of negative input voltage, while diode D2 is reverse biased and therefore capacitor C2 is disconnected from the rest of the circuit. In this stage, capacitor C1 charges to a DC voltage ideally equal to the amplitude of the AC signal. Afterward, during the phase of positive input voltage, the biasing of the diodes reverses: D2 is forward biased, and D1 becomes reverse biased. This causes capacitor C2 to be directly connected to the input of the AC signal. Taking into consideration the voltage at which capacitor C1 was originally charged, the DC voltage at capacitor C2 will be equal to twice the amplitude of

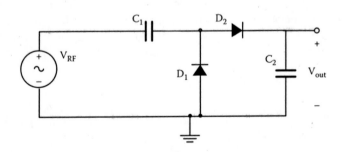

FIGURE 3.5
Structure of a single-stage basic voltage multiplier. V_{RF} models the voltage generated by the electromagnetic field in which the transponder is immersed.

the input voltage. Considering now the minimum ON voltage for real diodes, the voltage at the output of the structure shown in Figure 3.5 is

$$V_{out} = 2(V_p - V_{ON}) \tag{3.1}$$

where
V_{out} = the output DC voltage,
Vp = the peak voltage detected at the antenna terminals
V_{ON} = the minimum direct voltage across the diode to turn it on

Figure 3.6 shows the intermediate signals observed at the different nodes in the voltage doubler structure. The voltage supply models the radiofrequency signal detected at the terminals of the antenna in the transponder. The graph of the output voltages shows how the output capacitor is being charged until it reaches the steady-state level predicted by equation (3.1). Once the transient disappears, the output voltage stabilizes and will remain at that level as long as the transponder is immersed in that electromagnetic field.

In practical RFID systems, because the RF voltage detected at the antenna terminals is relatively small, the voltage doubler structure shown in Figure 3.5 does not produce a high enough voltage to power the integrated circuit in the transponder. This situation can be resolved by connecting additional voltage doublers. The resulting structure is known as a *Dickson charge pump*, resulting in an output DC voltage equal to

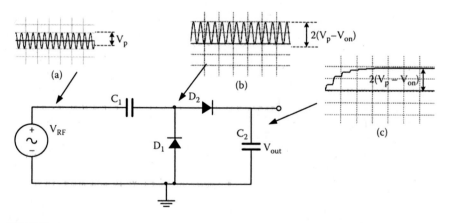

FIGURE 3.6
Signals observed at diverse points in the single-stage voltage multiplier. (a) Input signal. (b) Intermediate signal consisting of the input signal plus a DC component equal to its peak value. (c) Output signal. After the transient disappears, the output signal is a DC signal with a value equal to the double of the peak amplitude of the input signal.

$$V_{out} = 2N(V_p - V_{ON}) \qquad (3.2)$$

where N is the number of basic voltage doubler cells.

Figure 3.7 shows a three-stage Dickson charge pump. Its output voltage is equal to six times the peak voltage for the input minus the turn-on voltage for the diodes. Figure 3.8 shows the intermediate signals observed at several nodes in this structure. The sensitivity of the different graphs in volts per division has been kept the same to better observe how at each intermediate node, the AC signal has an additional offset equal to its peak value. The final DC output voltage is predicted by equation (3.2).

There is, however, a practical limit to the number of stages that is reasonable to consider for use in a practical application. As the number of stages increases, the number of required diodes increases by a factor of 2, and therefore the power dissipated by the charge pump also increases in the same ratio, decreasing the overall efficiency of the circuit. The efficiency of the charge pump is given by equation (3.3):

$$\eta_{pump} = \frac{P_{load}}{P_{load} + P_{consumed}} = \frac{V_{out}}{V_{out} + 2N V_{ON}} \qquad (3.3)$$

where N is the number of stages in the charge pump.

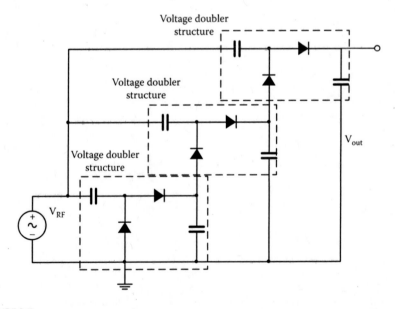

FIGURE 3.7
Dickson charge pump structure used to increase the DC voltage extracted from the radiofrequency signal generated by the electromagnetic field in which the transponder is immersed.

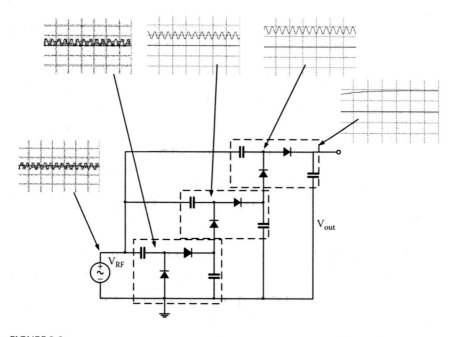

FIGURE 3.8
Signals observed at different points in the Dickson charge pump. After each stage, the overall signal increases its DC component until it only contains a DC component at the output of the last stage.

Equation (3.3) shows that the efficiency of the charge pump will increase by using diodes with a lower turn-on voltage. To this extent, several types of devices have been used and several more types are being currently investigated. The most common approach is to use Schottky diodes due to their electrical characteristics that make them extremely attractive for this purpose. However, using Schottky diodes increases the complexity of manufacturing the transponder, thus resulting in a higher cost. For this reason, the use of ultra-low-power complementary metal oxide semiconductor (CMOS) diodes appears to be an attractive alternative as they can be manufactured with the rest of the transponder and therefore they will not increase the cost of manufacturing the transponder.

3.2 Signals in the Transponder

3.2.1 Signal Encoding

The digital signals in the transponder are encoded using one or more of these common encoding schemes that are represented in Figure 3.9.

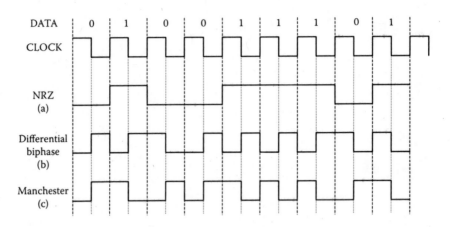

FIGURE 3.9
Common signal-encoding techniques. (a) Nonreturn-to-zero (NRZ): *1* represented by a high logic level; *0* represented by a low logic level. (b) Differential biphase: *1* represented by a change in level at the start of the clock period; *0* represented by no change in level at the start of the clock period. There is always a change of level at the middle of the clock period. (c) Manchester: *1* represented by a high-to-low transition at the middle of the clock period; *0* represented by a low-to-high transition at the middle of the clock period.

These encoding techniques are as follows:

NRZ (nonreturn to zero): There is no data encoding done. The 1's and 0's are clocked from the data array directly to the output transistor that will close or open the switch in the analog front end, as was shown in Figure 3.4. Although extremely simple, this encoding technique presents problems if the data to transmit have long strings of 1's or 0's because these may be misunderstood as the presence of a DC voltage in the line.

Differential biphase: Multiple encoding approaches using this method. This method embeds clocking information with the data and will help to synchronize the interrogator to the bit stream. In this approach, a level change occurs at the middle of every bit clock period. *1* is represented by a change in level at the start of the clock. *0* is represented by no change in level at the start of the clock.

Manchester code (biphase level, biphase_L, or split phase): This is a variation of the previous encoding method. There is not always a transition at the clock edge, but there is always a transition in the middle of the clock cycle, thus allowing one to extract the clock signal from the data signal. In Manchester code, a 1 is represented by a high to low level change in the middle of the clock. A *0* is represented by a low to high level change at the start of the clock. This type of encoding is used in the MCRF355/360 and the MCRF45X family of transponders manufactured by Microchip Technology.

Example 3.2:

Assume that we are using a biphase-L format (Manchester) for encoding the digital data at a frequency of 70 kHz. The data bit *0* is sent by first detuning (cloaking) the device during 7 μs and then tuning (uncloaking) the device for an additional 7 μs. The data bit *1* is sent by uncloaking the device and then cloaking it for the same time periods.

3.2.2 Modulation and Demodulation

Modulation refers to changing one or more parameters in the carrier radio-frequency signal as a method to transmit the encoded signal. In practice, only one of the parameters (amplitude, frequency, or phase) of the radiofrequency is usually changed when transmitting simple digital data.

When the amplitude of the carrier changes depending on the encoded signal, the resulting modulated signal is called *amplitude shift keying,* or ASK. In this case, a high voltage in the envelope of the radiofrequency signal signifies a logic level 1, while a low voltage in the envelope signifies a logic level 0, following the appropriate signal-encoding method being used. ASK modulation is also known as *direct modulation* in RFID systems. ASK modulation offers the possibility of high data rates due to the simplicity of the process. The spectrum of the ASK signal is relatively narrow as it contains only energy at the frequency of the carrier signal and at each of the two sidebands. One sideband is located at the frequency of the carrier minus the frequency of the modulating signal, and the other sideband is located at the frequency of the carrier plus the frequency of the modulating signal. The bandwidth efficiency of a binary-modulated ASK signal is 1 bit/second/Hz. The detection of the ASK signal can be done using a coherent detector or a noncoherent detector. Noncoherent detection increases the simplicity of the overall system, although it reduces the ability to differentiate between the desired signal and noise. The immunity to noise can be increased by using a coherent detector, although the immunity to noise is overall lower than using an FSK or PSK modulation process. The bit error probability for noncoherent ASK modulation, assuming an additive white Gaussian noise limited channel, is

$$P\Big|_{\substack{ASK \\ NONCOH}} = 0.5e^{\frac{-E_b}{4N_0}} + 0.5erfc\left(\sqrt{E_b/2N_0}\right) \qquad (3.4)$$

where
P = the bit error probability
E_b/N_0 = the average bit power
$erfc(x)$ = the complementary error function for the argument x

Similarly, the bit error probability for coherent ASK modulation is

$$P|_{\substack{ASK \\ COH}} = 0.5 erfc\left(\sqrt{E_b / 2N_0}\right) \tag{3.5}$$

Figure 3.10 shows the bit error probability values obtained using equations (3.4) and (3.5) for different values of energy transmitted. From this graph it is possible to observe the decrease in bit error rates for coherent ASK modulation, which is especially significant when increasing the energy per bit transmitted.

When the encoding signal changes the amplitude of the carrier being transmitted, the resulting modulated signal is called *frequency shift keying*, or FSK. When the modulating signal is binary, FSK results in using two different frequencies for transmitting the digital data. The most common approach to FSK in RFID systems is called Fc/8/10: this means that a *0* is transmitted as an amplitude modulated clock cycle with a period corresponding to the carrier frequency divided by 8, and a *1* uses a period that corresponds to the carrier frequency divided by 10. Figure 3.11 shows the time representation for this FSK modulation.

The logic level *0* is transmitted by sending a total of eight cycles of the RF signal; the first four cycles have a higher amplitude than the last four cycles. The logic level *1* is transmitted by sending a total of ten RF cycles,

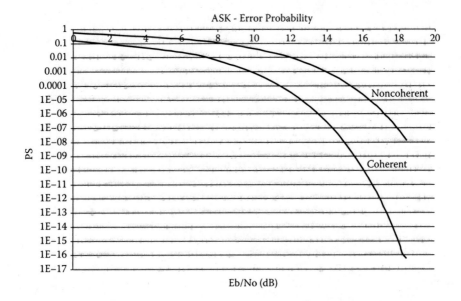

FIGURE 3.10
Error probability (Ps) for coherent and noncoherent ASK modulation as a function of the energy received (Eb/No).

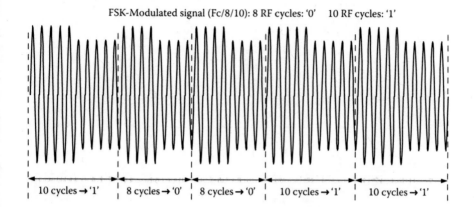

FSK-Modulated signal (Fc/8/10): 8 RF cycles: '0'   10 RF cycles: '1'

| 10 cycles → '1' | 8 cycles → '0' | 8 cycles → '0' | 10 cycles → '1' | 10 cycles → '1' |

FIGURE 3.11

Structure of the FSK-modulated signal. The symbol *1* is represented by sending 10 RF cycles, while the symbol *0* is represented by sending 8 RF cycles. The message sent in the figure is *10011*.

with the first five cycles having a higher amplitude than the last five cycles. The change in amplitude between the first and second halves of the cycles allows one to count the number of cycles between transitions and therefore to differentiate between logic levels. In other words, the *1* and *0* are differentiated by looking at the number of periods between transitions.

Other FSK approaches used by other transponders are Fc/10/8, Fc/5/8, and Fc/8/5. Table 3.1 lists the relation between the encoding signal and the data transmitted.

The bandwidth required by the FSK signal is dependent on the existence or not of phase changes when changing frequencies. Obviously, in those FSK modulation approaches that exhibit abrupt phase changes, there will be more spectral components at higher frequencies, thus increasing the bandwidth. The bandwidth efficiency of a binary-modulated FSK signal depends on how the modulation process has been performed, but it approaches 2 bits/second/Hz. Similarly to ASK modulation, FSK signals can be recovered using noncoherent detectors or coherent detectors, with a coherent detector providing stronger noise immunity. The calculation of the bit error rate for FSK signals depends on the separation between the frequencies used to transmit the FSK signal. The optimal frequency selection occurs when these frequencies are a multiple of the frequency of the encoding signal. These are called *orthogonal frequencies*.

The bit error probability for noncoherent FSK modulation, assuming an additive white Gaussian noise limited channel, is

$$P\big|_{\substack{FSK \\ NONCOH}} = 0.5e^{\frac{-E_b}{2N_0}} \tag{3.6}$$

TABLE 3.1

Different FSK Modulation Approaches

	Logic Level "0"	Logic Level "1"
Fc/8/10	8 RF cycles	10 RF cycles
Fc/10/8	10 RF cycles	8 RF cycles
Fc/5/8	5 RF cycles	8 RF cycles
Fc/8/5	8 RF cycles	5 RF cycles

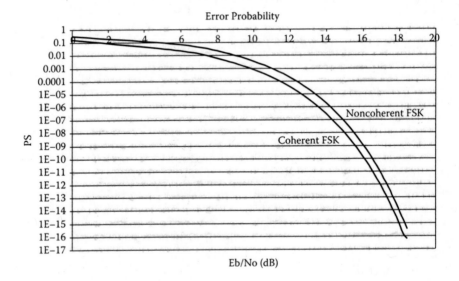

FIGURE 3.12

Error probability (Ps) for noncoherent FSK and coherent FSK as a function of the energy received (Eb/No).

The bit error probability for coherent FSK modulation is

$$P\big|_{\substack{FSK \\ COH}} = 0.5\,erfc\left(\sqrt{E_b / 2N_0}\right) \tag{3.7}$$

Comparing equations (3.5) and (3.7) shows that if the two frequencies that make the FSK signal are chosen to be orthogonal, then the bit error rates for coherent ASK and coherent FSK signals are essentially the same. Figure 3.12 shows the probability of bit error as predicted by equations (3.3) and (3.4) for different values of energy per bit. This figure shows that the difference in performance between noncoherent detection and coherent detection for FSK signals is lower than for ASK signals. This has resulted in the widespread use of noncoherent FSK modulation due to its simplicity.

Example 3.3:

Calculate the bit error rate for coherent and noncoherent ASK and FSK demodulation for the following signal-to-noise ratios (E_b/N_o): (a) 5 dB, (b) 10 dB, and (c) 15 dB.

Solution:

The bit error rates for each type of demodulation are given by equations (3.4) to (3.7). The first step is to express the signal-to-noise ratio in linear units. This results in the following values:
   A. E_b/N_o = 5 dB ➔ E_b/N_o = 3.16 (linear units)
   B. E_b/N_o = 10 dB ➔ E_b/N_o = 10.0 (linear units)
   C. E_b/N_o = 15 dB ➔ E_b/N_o = 31.6 (linear units)

Bit error rates for noncoherent ASK demodulation are predicted by equation (3.4), resulting in the following:
   With E_b/N_o = 3.16 (5 dB): P_s = 0.264 (approximately 1 error every 4 bits transmitted)
   With E_b/N_o = 10.0 (10 dB): P_s = 0.042 (approximately 1 error every 24 bits transmitted)
   With E_b/N_o = 31.6 (15 dB): P_s = 0.185 10^{-3} (approximately 1 error every 5400 bits transmitted)

Bit error rates for coherent ASK demodulation are predicted by equation (3.5), resulting in the following:
   With E_b/N_o = 3.16 (5 dB): P_s = 0.037 (approximately 1 error every 30 bits transmitted)
   With E_b/N_o = 10.0 (10 dB): P_s = 0.78 10^{-3} (approximately 1 error every 1300 bits transmitted)
   With E_b/N_o = 31.6 (15 dB): P_s = 9.47 10^{-9} (approximately 1 error every 100 million bits transmitted)

Bit error rates for noncoherent FSK demodulation are predicted by equation (3.6), resulting in the following:
   With E_b/N_o = 3.16 (5 dB): P_s = 0.103 (approximately 1 error every 10 bits transmitted)
   With E_b/N_o = 10.0 (10 dB): P_s = 3.37 10^{-3} (approximately 1 error every 300 bits transmitted)
   With E_b/N_o = 31.6 (15 dB): P_s = 6.87 10^{-8} (approximately 1 error every 14.5 million bits transmitted)

Bit error rates for coherent FSK demodulation are predicted by equation (3.7), resulting in the following:
   With E_b/N_o = 3.16 (5 dB): P_s = 0.037 (approximately 1 error every 30 bits transmitted)
   With E_b/N_o = 10.0 (10 dB): P_s = 0.78 10^{-3} (approximately 1 error every 1300 bits transmitted)
   With E_b/N_o = 31.6 (15 dB): P_s = 9.47 10^{-9} (approximately 1 error every 100 million bits transmitted)

These results show that the performance of coherent demodulation has a much better performance than noncoherent demodulation. Even for relatively noisy signals, coherent ASK shows a much lower bit error rate than when the demodulation is performed noncoherently. Therefore, it seems obvious to choose a coherent demodulation scheme when using ASK. While coherent FSK demodulation also shows a lower bit error rate than noncoherent FSK demodulation, the difference is not as strong as in the ASK case. For this reason, a noncoherent FSK demodulation may be preferred as it will reduce the complexity of the system without excessively degrading its performance.

When the encoding signal changes the phase of the carrier, the resulting modulated signal is called *phase shift keying* (PSK), as shown in Figure 3.13. PSK is basically the same approach as FSK with the difference being that only one frequency is used. The information is encoded here in the change of phase between clock cycles. PSK also allows for different approaches on this encoding. Because the receiver is only looking for changes of phase, it is possible to transmit information in a smaller number of radiofrequency cycles, resulting in using faster data transfer rates compared to FSK.

Some devices like the MCRF200 and the MCRF250, both from Microchip Technology, support two PSK modulation modes called PSK_1 and PSK_2. In PSK_1, the phase of the RF signal changes when the encoding data change. This means that, for example, the phase of the RF signal will change when the encoding signal changes from 1 to 0 or from 0 to 1. In

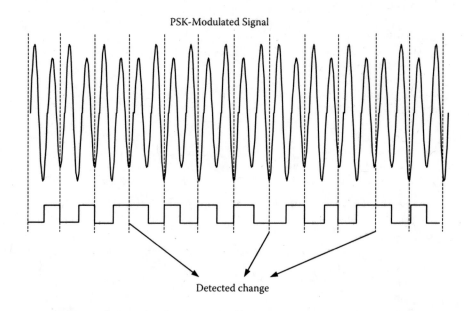

FIGURE 3.13
A generic PSK-modulated signal. The figure shows that the PSK signal requires less RF cycles to transmit the digital data.

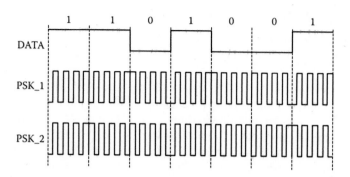

FIGURE 3.14

Two variations of PSK modulation. In PSK_1, the phase of the carrier changes every time there is a change in the symbol transmitted. In PSK_2, the phase of the carrier changes every time the symbol transmitted is 1.

PSK_2, the phase of the RF signal changes every time the encoding data have the logic level 1. For example, the phase will change when the encoding data change from 0 to 1 or from 1 to 1, but it will not change when the encoding data move from 1 to 0 or from 0 to 0. Figure 3.14 depicts these two PSK modulation approaches.

A third PSK modulation mode available in some other devices—for example, the T5557 from Atmel®—is based on a phase change on the rising edge of the encoding signal.

The spectral occupancy of the PSK signal is identical to the ASK signal assuming that there are no phase changes abruptly occurring at the symbol boundaries. However, in contrast to ASK and FSK, the detection of PSK signals requires a coherent demodulator. The bit error rate for PSK signals is

$$P\big|_{\substack{PSK \\ COH}} = 0.5 \, erfc\left(\sqrt{E_b / N_0}\right) \tag{3.8}$$

When comparing the different modulation methods and assuming equal average energy per symbol transmitted, the different PSK modulation approaches present the strongest immunity against noise. Coherent FSK modulation with orthogonal symbols or coherent ASK modulation is the next best approach followed by the noncoherent FSK modulation. Noncoherent ASK modulation exhibits the poorest performance. For example, a bit error rate of 10^{-6} requires a value of Eb/No slightly above 10 dB for a PSK signal, but it will require a value of Eb/No of about 18 dB for a noncoherent ASK signal. Similarly, an energy Eb/No of 10 dB will result in a bit error rate of $1.5 \ 10^{-5}$ in using PSK, but the bit error rate will increase up to $1.5 \ 10^{-2}$ when using noncoherent ASK. This represents moving from 1.5 errors every 100,000 bits transmitted (PSK) to 1.5 errors every 100 bits transmitted (noncoherent ASK).

3.3 The Logic System

3.3.1 Data Memory

The memory of RFID transponders is divided into data memory and configuration memory. Data memory stores the information that will be transmitted back to the interrogator. Configuration memory stores data regarding the configuration parameters for the transponder. Memory varies in size, rewriting capabilities, structure, and characteristics similar to those found in memory used in other products. The size of the memory ranges from 1 bit to 64 kbits or even larger, with the most common values being around a few hundred bits. One of the most important differences between transponders is whether the data stored in their memory can be rewritten by the user. This gives rise to the distinction between *read-only* transponders and *read/write* transponders. Because memory types, sizes, and structures can greatly differ between different transponders, the following examples are not intended to be viewed as a comprehensive guide, but to highlight some of the most commonly used approaches to memory in transponders.

The device MCRF200 from Microchip Technology contains a memory of 140 bits; 128 bits are data bits, and the remaining 12 bits are used for the configuration register. This is a one-time programmable (OTP) device that operates as a read-only device once it is programmed. The data memory can be programmed by the user using a contactless programmer or can be directly programmed by the factory at the time of production. The configuration register can only be programmed at the time of production. The manufacturer specifies memory data retention better than 200 years.

Figure 3.15 shows the block diagram of this device. The EEPROM memory is addressed by the column and row decoders at the clock rate. The output from the memory array is a bit stream that is directly fed into the modulation control and modulation circuit in order to be transmitted back to the interrogator. Figure 3.16 shows the description of the different bits for the configuration register in this memory.

CB12 is the bit that allows programming the device. Once CB12 is set to 1, the device cannot be programmed or erased. Because the MCRF200 does not support anticollision, bit CB11 is always 0. However, for the device MCRF250 that is similar to the MCRF200 with the only exception that MCRF250 supports anticollision, its bit CB11 is always set to 1. CB10 is used to establish the transmission rate when PSK modulation has been selected. Bits CB9 and CB8 establish the modulation for the encoded data using the modulation types shown in Section 3.2.2. Bits CB7 and CB6 determine how the data are encoded, as described in Section 3.2.1. Bit CB5 is always set to 0. Bits CB4, CB3, and CB2 are used to set the baud rate at which data are transferred. The default timing (CB4 = CB3 = CB2 = 0) is MOD128, meaning that the transmission rate is set at 128 RF cycles for bit.

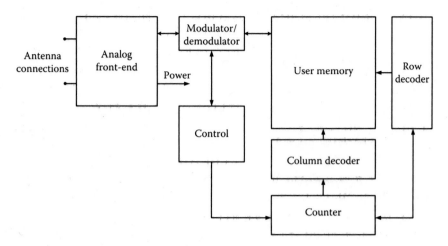

FIGURE 3.15
Structure of the circuitry for memory reading, writing, and control.

Because the MCRF250/251 device operates in the LF range, assuming a carrier frequency of 128 kHz, this setting will result in a transmission rate of 1 kHz. Conversely, the fastest transmission rate is MOD16, which will result in a transmission rate of 8 kHz. It is also important to keep in mind that MOD128 is the data rate used when programming the device. Finally, CB1 is used to specify the size of the data memory of 128 bits or 96 bits depending upon the needs of the user.

Example 3.4:

The configuration register for a MCRF 200 device is set to $08D. Describe the settings for the device.

Solution:

The binary values for the configuration register are shown as follows:

CB12	CB11	CB10	CB9	CB8	CB7	CB6	CB5	CB4	CB3	CB2	CB1
0	0	0	0	1	0	0	0	1	1	0	1

CB12 = 0. The device has not been programmed yet.
CB11 = 0. MCRF200 does not support anticollision; therefore, CB11 is set to 0.
CB10 = 0. Rate of PSK modulation is equal to half of the carrier frequency.
CB9 = 0 and CB8 = 1. The modulation chosen is PSK_1. This means that the phase of the modulation signal changes at the change of the encoded data.
CB7 = CB6 = 0. Data encoded using nonreturn to zero level (NRZ_L).
CB5 = 0. No special meaning because CB5 is always set to 0.
CB4 = 1; CB3 = 1; CB2 = 0. Baud rate is MOD32, resulting in a 4 kHz baud rate.
CB1 = 0. Memory size is 128 bits.

CB12	CB11	CB10	CB9	CB8	CB7	CB6	CB5	CB4	CB3	CB2	CB1
Lock	Anticol.	PSK	Modulation		Encoding		Sync	Baud Rate			Size

CB12: Memory array lock bit
0: User memory not locked
1: User memory locked

CB11: Anti-collision option
0: Disabled

CB10: PSK Rate
0: Carrier/2
1: Carrier/4

CB9, CB8: Modulation

CB 9	CB 8	Modulation
0	0	FSK (fig. 3.11)
0	1	PSK_1 (fig. 3.14)
1	0	ASK
1	1	PSK_2 (fig. 3.14)

CB7, CB6: Encoding

CB 7	CB 6	Encoding (fig 3.9)
0	0	NRZ
0	1	Manchester
1	0	Diff. Biphase
1	1	Inverted manchester

CB5: Sync
CB5 = 0 always

CB4, CB3, CB2: Baud rate

CB 4	CB 3	CB 2	Rate
0	0	0	MOD 128
0	0	1	MOD 100
0	1	0	MOD 80
0	1	1	MOD 32
1	0	0	MOD 64
1	0	1	MOD 50
1	1	0	MOD 40
1	1	1	MOD 16

CB1: User memory size
0: 96 bits
1: 128 bits

FIGURE 3.16
Configuration bits used in the MCRF200 transponder.

Example 3.5:

State the configuration register for a transponder based on the MCRF250 device with the following parameters:
   Blank (not programmed device)
   Modulation: PSK_1.
   PSK rate: Frequency half of the carrier frequency
   Data encoding: NRZ_L
   Baud rate: MOD 32
   Memory size: 128 bits

Solution:

The MCRF250 is a device similar to the MCRF200, with the only difference being that it supports anticollision. Therefore, the configuration bit CB11 will be set to 1. The rest of the configuration bits have the same meaning as the MCRF200. The values of the configuration bits in this case are as follows:

CB12	CB11	CB10	CB9	CB8	CB7	CB6	CB5	CB4	CB3	CB2	CB1
0	1	0	0	1	0	0	0	1	1	0	1

Therefore, the hexadecimal value stored in the configuration register is $48D.

The devices MCRF355 and MCRF360, also manufactured by Microchip Technology, operate at the HF frequency of 13.56 MHz. They both have a total of 154 bits of memory that must be programmed using a contact programmer. This makes them act as read-only devices when they are used in the field but gives them the flexibility to be reprogrammed by the user when necessary. The data retention specified by the manufacturer is also better than 200 years.

The family of devices MCRF450/451/452/455 also operates in the HF frequency range at 13.56 MHz. As was discussed in Chapter 2, the different devices in this family use internal or external capacitors to modulate the radiofrequency signal. In discussing their internal memory structure, the concepts described for the device MCRF450 are also valid for the rest of the devices in this family. All of these devices have their memory organized in two areas: main memory and stored cyclic redundancy check (SCRC) memory. SCRC memory is used in the operation of the cyclic redundancy code necessary for the operation of the anticollision algorithm, as all of these devices support anticollision. The explanation of the CRC procedure is shown in Section 3.3.2.

The main memory of the MCRF450 and similar devices contains a total of 1024 bits and is structured in 32 blocks, each one with 32 bits. The SCRC memory is organized in 32 blocks, each one with 16 bits. Figure 3.17 shows the structure of the user memory in these devices.

The first three blocks (B0–B2) of the main memory are used to set up the operation of the device in a manner similar to a configuration register, while the remaining 29 blocks (B3–B31) are for user data. All the blocks of the main

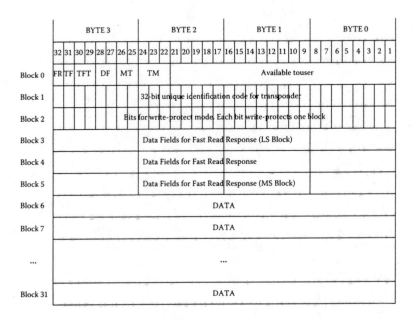

FIGURE 3.17
Structure of the user memory for the MCRF 450 family of transponders.

memory with the exception of Block 1, which contains the ID or serial number of the transponder, can be individually rewritten block by block by a contactless programmer. Block 1 is programmed at the factory and is protected against being rewritten. The main memory is read or written in blocks of 32 bits, with the exception of bits 30 and 31 in Block 0 that can be selected individually. Blocks 3 to 5 constitute the fast read field (FRF) that is used by the interrogator in the transmission of commands to the transponder.

Block 0 is split into 21 bits of general memory available to the user and 11 bits containing operational parameters for the transponder. These operational bits, shown in Figure 3.17, are as follows:

FR (Bit 31). When FR = 0, the transponder will respond to the *fast read bypass* (FRB) command but will not respond to the *fast read request* (FRR) command from the interrogator. When FR = 1, the previous situation is reversed. These interrogator commands are further explained and described in Chapter 6.

TF (Bit 30). TF = 0 sets the transponder in *interrogator talks first* (ITF) mode. This means that it will wait for an FRR command. TF = 1 sets the transponder in *tag talks first* (TTF) mode if FR = 1. In this case, the transponder will send a *fast read response* without waiting for an FRR.

TFT (Bits 29–28). These two bits set the value for the parameter TCMAX when the transponder is in TTF mode. TCMAX is the number of

TABLE 3.2

Configurations for TFT Bits

Bit 29	Bit 28	TCMAX
0	0	1
0	1	2
1	0	4
1	1	Continuous

TABLE 3.3

Configuration for DF Bits

Bit 27	Bit 26	TCMAX
0	0	32 Bits
0	1	48 Bits
1	0	64 Bits
1	1	96 Bits

fast read responses that the transponder can send after an FRR command, as shown in Table 3.2. For example, TCMAX = 2 means that the transponder can send its response twice for acknowledgment. If TCMAX is set to continuous mode, the transponder will send its response approximately every 80 ms until it receives the correct response from the interrogator. This is the default mode.

DF (Bits 27–26). These two bits set the length of the data transmitted by the transponder according to Table 3.3. The default data length is 32 bits.

MT (Bits 25–24). These two bits are hardwired and cannot be changed by the user. At the current state of production, these two bits are both set to 0. The remaining values are reserved for future applications.

TM (Bits 23–22–21). These three bits set up the total memory size. These three bits are also hardwired. Currently, there are only two possibilities allowed: 000 and 001. The code *000* describes a transponder with 512 bits of memory, while the code *001* describes a transponder with a memory size of 1 kbit.

Block 1 contains a unique 32-bit identification number for the transponder. This identification word is serialized by the manufacturer.

Block 2 contains the bits that set each block in write-protect mode. Each bit corresponds to its 32-bit block. For example, bit 5 will set Block 5 as writable or write-protected. When the write-protect bit is set to 1, the block is writable; when the bit is set to 0, the block is write-protected. It is important to note that once a block is write-protected, it cannot be changed back to being writable. The transponders are shipped to the user with all the write-protect bits to 1 with the exception of bit 1. This means that Block 1, which contains the unique ID for the transponder, cannot be changed by the user. Bits 31 and 30 in Block 0 (bits FR and TF) are not write-protectable, thus allowing the user to change how the transponder is operating. Finally, it is also important to note that Block 2 can also be set to be write-protected.

Blocks 3 to 5 contain data bits for the fast response. The size of the fast response was set by bits 27 and 26 (bits DF) in Block 0.

The SCRC memory section is organized into 32 blocks, each one with 16 bits as shown in Figure 3.18. The blocks contain the CRC code for the corresponding memory block sent back to the interrogator in order to check for anticollision.

One of the family of transponders operating at 13.56 MHz manufactured by Atmel® has an internal user memory ranging from 1 kbit to 64 kbits with the distinct feature that this memory is encrypted, thus providing an additional layer of protection. The *CryptoRF EEPROM* family offers user memory of 1 kbit, 2 kbits, 4 kbits, 8 kbits, 16 kbits, 32 kbits, and 64 kbits. In addition to the user memory, all these devices have a configuration memory of 2 kbits. This configuration memory is used to store eight sets of passwords for reading and writing, four crypto key sets, security access registers for each user zone and password, and key registers for each zone. The transponders in this family also support anticollision. Figure 3.19 shows the block diagram for these types of devices. As shown in the figure, while the analog front end is similar to that of any transponder, it differs in the existence of authentication and password verification prior to the device sending data.

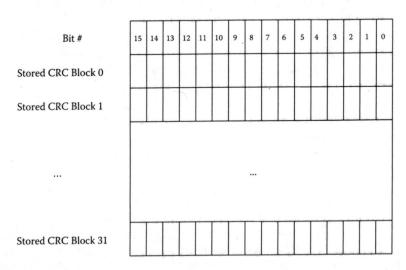

FIGURE 3.18
Structure of the stored CRC memory used in the MCRF 450 family of transponders.

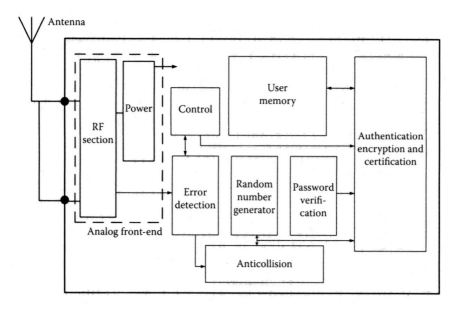

FIGURE 3.19
Transponder with encrypted memory showing the additional components needed for this process.

In the transponders with user memory size of 64 kbits, 32 kbits, or 16 kbits, the EEPROM user memory is divided into 16 user zones. Figure 3.20 shows the memory map for the 64 kbit memory transponder. As seen in the figure, each user zone is configured for storing 512 bytes (4096 bits), giving a total user memory of 64 kbits. For the 32 kbit transponder, its memory is also divided into 16 user zones, each one of 256 bytes (2048 bits); for the 16 kbit memory, the size of each user zone is 128 bytes (1024 bits).

The transponder from the same family with a memory of 8 kbits has a similar structure, although the memory is divided into eight different user zones. Each zone can store 128 bytes (1024 bits). The memory for the transponders with 4 kbits, 2 kbits, and 1 kbit of user memory are all divided into four user zones. For the 4 kbit memory, each user zone stores 128 bytes (1024 bits); for the 2 kbit memory, each user zone stores 64 bytes (512 bits); and for the 1 kbit memory, each user zones stores 32 bytes (256 bits).

In all cases, the access to the user zones is only possible after having some security requirements are met. These are defined by the user during the setup of the device. The initial state for the transponders in this family is to have all the security features disabled. The user must set up the level of protection by configuring the security requirements for the transponder. These security requirements are stored in the configuration memory.

All the transponders in the CryptoRF family have a configuration memory of 2048 bits. This memory stores passwords, keys, codes, and security

	Address	Bit 7	Bit 6	Bit 5	Bit 4	Bit 3	Bit 2	Bit 1	Bit 0
User Zone 0	0								
	1								
	...	Total of 512 bytes (4096 bits) in Zone 0							
	512								
User Zone 1	0								
	1								
	...	Total of 512 bytes (4096 bits) in Zone 1							
	512								
User Zones 2 to 14	...								
User Zone 15	0								
	1								
	...	Total of 512 bytes (4096 bits) in Zone 15							
	512								

FIGURE 3.20
Memory map divided into 16 user zones, each one containing 4096 bits for a total of 65,536 bits (64 kbits).

definitions for each one of the user zones. The access rights to the configuration memory are defined in the control logic and cannot be altered by the user. The transponder can be programmed so each zone uses a different set of passwords. This is especially useful in the case of different users accessing the transponder. In addition, the transponders include three fuses that must be blown during the setup process. The fuses lock some of the portions of the configuration memory.

Atmel® also manufactures families of transponders that operate in the LF range, as well as in the UHF range. For example, the transponder ATA5558 can operate between 100 kHz and 200 kHz, while the ATA5590 operates in the UHF range from 860 MHz to 960 MHz. Both devices contain an internal EEPROM divided into a user memory of 1024 bits and a system memory of 320 bits. The user memory is organized in pages of 128 bits, and each page in blocks of 32 bits. Each one of the blocks must be programmed separately. Pages are protected against overwriting by the use of a lock bit that, as shown in Figure 3.21, is the most significant bit of each block.

The system memory in the ATA5590 transponder contains a page with the identification information for the transponder (Tag_ID page), a page with system-level information available to the user, and two blocks with system-level information for the manufacturer of the transponder.

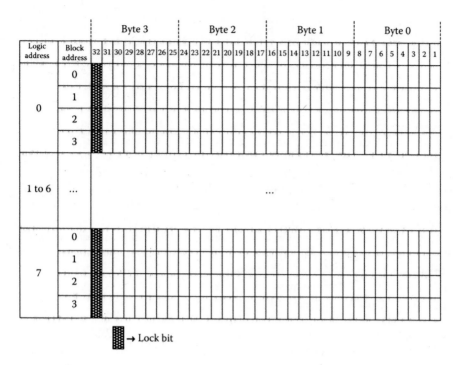

→ Lock bit

FIGURE 3.21
Memory map with lock bit. The lock bit is located at the MSB in each address.

3.3.2 Cyclic Redundancy Check for Error Detection

The cyclic redundancy check is the most used method for ensuring the integrity of the digital signals in some RFID transponders. In particular, CRC is used as a method of error detection in the digital stream transmitted and received by the transponders that support CRC. It is important to note that although the CRC algorithms detect the existence of errors, they do not make corrections.

The CRC method is based on appending additional information to the message being transmitted. This additional information is mathematically related to the message and therefore is redundant. At the receiving end, the receiver checks this additional information to verify that it agrees with the message being transmitted. This allows the receiver to determine with a certain degree of probability the existence of an error in the transmission. However, this method implies that not all errors are subject to detection. Simply stated, the CRC method is implemented by dividing the entire numeric binary value of the message by a constant. This constant is called a *generator polynomial*. The reminder in the division is then appended to the message. The polynomial that is used for the division is chosen from a family of polynomials with certain mathematic properties whose study is beyond the scope of this book. The transponders from Atmel and Microchip both use a polynomial known as the CRC-CCITT-16, although their specific implementation is somewhat different. This polynomial is also used in other digital transmissions such as CDMA and Bluetooth, among others. The CRC-CCITT-16 polynomial is

$$CRC - CCITT - 16 : x^{16} + x^{12} + x^5 + x^0$$

Reading the polynomial from most significant bit (MSB) to least significant bit (LSB) (normal representation) results in a value of 1021 hexadecimals. We must take into consideration that CRC polynomials have always had their MSB equal to 1, and for this reason it is not considered in the calculation of their values. Reading the polynomial from LSB to MSB (reverse representation) results in the value of 8048 hexadecimals. Microchip Technology uses reverse representation, while Atmel® uses normal representation.

CRC can be hardware-implemented using shift registers and exclusive-OR (XOR) gates in the feedback process shown in Figure 3.22. The encoder shown in Figure 3.22 consists of 16 shift registers and the exclusive-OR gates. The initial value loaded into the shift registers is $FFF in the Microchip Technology approach and $000 for the Atmel approach. The encoder performs the exclusive-OR function and shifts the registers until the last bit of the data stream is entered. At that point, the CRC value of the data set is equal to the values in the shift registers. When the transponder transmits data, the calculated CRC value is attached to the data.

The receiver performs an identical process and verifies that the CRC code generated by the receiver is equal to the CRC code appended to the message. The advantage of the CRC method is that the current state of one of the shift registers is a result of considerable past history, as it has to go through a lengthy process. Therefore, it is unlikely that a burst of errors will produce a CRC code equal to the one generated in the absence of errors. In fact, the CRC-CCIT-16 code is able to detect all error bursts of 16 bits or less as well as 99.9% of error bursts of more than 16 bits.

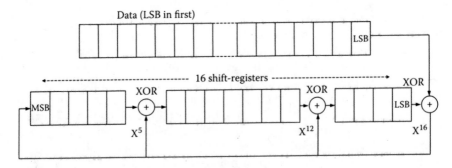

FIGURE 3.22
Structure of a generic CRC encoder using the CRC-CCITT-16 polynomial.

Different manufacturers may also follow different approaches as to whether they append the CRC value with its LSB or MSB first. The specification documents for the transponders describe how the CRC value is attached to the data. The previous section described the structure of memory in some specific transponders. For example, the user memory for the MCRF450 family of transponders from Microchip Technology is divided into 32 blocks. When the interrogator tries to write data into any of these blocks, before doing any processing, the transponders check the CRC. If the CRC data are correct, the transponder proceeds with storing the data and the CRC in memory. Then, the transponder immediately transmits back the data and stored CRC to the interrogator for verification. If the CRC received by the transponder is incorrect, the transponder ignores the received message and waits for the next command with a valid CRC.

3.4 Transponder Programming

The programming of the transponders can be done by the manufacturer at the time of production or by the end user. *Programming the transponders* means to store a unique identification number in its memory, as well as to set up the configuration parameters for the communication between the transponder and the interrogator and additional parameters such as password protection. Transponders can be programmed using contact or noncontact methods.

The MCRF200/250 family of transponders manufactured by Microchip Technology is an example of a contactless programmable device. Because the MCRF200/250 family is made up of one-time programmable devices, the device can only be programmed when it is blank from the manufacturing process. Programming this device requires a specific sequence of radiofrequency signals applied following this sequence:

Initial, power-up RF signal (125 kHz for the MCRF200/250 family) lasting between 80 µs and 180 µs.

Absence of RF field lasting between 50 µs and 100 µs.

A continuous FSK signal that serves as a *verify* signal, lasting 131 ms. This signal is required for the transponder to be energized in order to output all the data contained in its 128 bits of memory. Because the device is blank at the time of programming, the output data are all *1*.

Programming data: After completing the verify mode, the device enters the programming mode. The device is programmed starting with bit 1 and finishing with bit 128. The symbol *1* is programmed by sending a low-power RF signal with an amplitude similar to the one used for the initial power-up. The symbol *0* is programmed by sending a high-power RF signal with an amplitude approximately equal to 2.2 times the amplitude used for the symbol *1*.

Device response: After the 128 bits have been programmed, the device responds by transmitting the programmed data.

The MCRF355/360 family of transponders, also from Microchip Technology, has 154 bits of memory that can be programmed by using a contact programmer. Also, the memory in this family of devices can be reprogrammed. Figure 3.23 shows the pin diagram for this device.

Note that Pin 3 (Ant. A) and Pin 6 (Ant. B) are used to connect the external capacitors and inductors, as was shown in Figures 2.16 and 2.17, while Pin 5 (Vss) serves a ground. In order to program these devices, Pins 3, 6, and 5 are connected to ground, Pin 8 (Vdd) is connected to the positive voltage source, Pin 2 (CLK) is connected to the clock, and Pin 1 (Vprg) is connected to the signal that will be programmed in the transponder.

Some specific codes are used to carry out specific functions in the memory of the device. For example, because this device is reprogrammable, when the user wants to write new data into memory, the existing data must be erased first. This is done by sending the code 0111010100 through the VPRG pin, as shown in Figure 3.24.

Similarly, to start the programming sequence, the user must transmit the code 0111010010, followed by the data bits to store in memory. In total, there will be a total of 166 bits sent to the device: the first 12 bits are the command to program, and the remaining 154 bits are the data to be stored in memory. Finally, the data stored in memory can be read by transmitting the code 0111010110 to the device. After this code has been transmitted, the device will output the data stored in its memory.

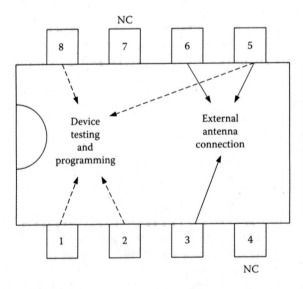

Pin	Function
1	VPRG
2	CLK
3	Antenna A
4	NC
5	V_{ss}
6	Antenna B
7	NC
8	V_{dd}

FIGURE 3.23
Pins for the MCRF355/360 transponders.

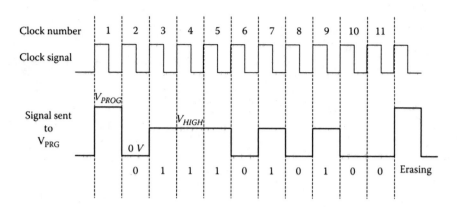

FIGURE 3.24
Code sent to erase existing data in memory using a contact programmer.

3.5 Summary of Electrical Specifications for RFID Transponders

The manner in which manufacturers specify the performance of their transponders varies greatly, not only among different manufacturers but also for devices from the same manufacturer. Tables 3.4 to 3.7 attempt to show the most important performance specifications from different manufacturers of RFID transponders. Table 3.4 does not pretend to be an exhaustive list of available devices, but just aims to show specific examples of device performance. The values of the parameters found in Tables 3.4 to 3.7 should be seen as somewhat typical performance values.

Table 3.4 shows the performance characteristics of three RFID transponders manufactured by Microchip Technology. The device MCRF200 is listed as a *125 kHz microID® Passive RFID Device*; the MCRF250 is exactly similar to the MCRF200 but incorporates anticollision; the MCRF355/360 family of devices is listed as *13.56 MHz Passive RFID Devices with Anti-Collision Feature*; and the devices in the MCRF450 family are listed as *13.56 MHz Read/Write Passive RFID Devices*.

Table 3.5 shows the electrical performance characteristics for four transponders manufactured by Atmel®. The ATA5567 device is listed as a *Multifunctional 330-Bit Read/Write RF Identification IC*; the ATA5570 is listed as a *Multifunctional 330-Bit Read/Write RF Sensor Identification IC*; the ATA 5558 is listed as a *1 kbit R/W IDID® with Deterministic Anticollision*; and the T5557 is listed as a *Multifunctional 330-Bit Read/Write RF Identification IC*.

Table 3.6 shows the performance parameters of two additional transponders from Atmel®. The device TK5530 is listed as a *read-only transponder,* while the TK5551 is listed as a *standards read/write ID transponder with anticollision.* The difference between these transponders and those evaluated earlier is

TABLE 3.4

Summary of Electrical Parameters for Several RFID Transponders Manufactured by Microchip Technology

Parameter	MCRF 200/250	MCRF 350/360	MCRF 450/451/452/455
Operating frequency	100 kHz to 400 kHz	13.56 MHz	2 MHz to 35 MHz
Dynamic coil current	50 μA		
Operating current	5 μA	7 μA	20 μA
Maximum current through antenna pads	50 mA	40 mA	40 mA
Current leakage		10 nA	
Turn-on voltage	10 Vpp between antenna pads 2 V_{DC}		
Reading voltage		2.4 V	2.8 V
Testing voltage		4.0 V	
Programming voltage		V_{IH}: 2.8 V V_{IL}: 1.2 V	
Coil voltage while reading		4 Vpp	4 Vpp
Coil clamp voltage		32 Vpp	
Programming time	2 sec (for all 128 bit array)		5 ms (for a 32-bit block)
Sleep time		50 ms to 200 ms	
Modulation resistance		< 4 Ω	< 5 Ω
Internal capacitance	2 pF	100 pF	95 pF (MCRF451) 30 pF (MCRF 452) 50 pF (MCRF 455)
Data retention	200 years	200 years	200 years
Maximum storage temperature	-65°C to +150°C	-65°C to +150°C	-65°C to +150°C
Maximum ambient temperature	-40°C to +125°C		-40°C to +125°C
Maximum dissipated power			500 mW

[a] Blank spaces in the table indicate that the manufacturer did not provide information for that specific parameter.

that the ones shown below incorporate the coil for the antenna within the transponder. Moreover, the operating characteristics are given as strength of magnetic field instead of voltage or current.

Table 3.7 summarizes the most important characteristics of several transponders manufactured by Texas Instruments under their Tag-it™ commercial name. The manufactured has also chosen to specify their reading and writing thresholds as field intensity.

TABLE 3.5

Summary of Electrical Parameters for Several RFID Transponders Manufactured by Atmel®

Parameter	ATA5567	ATA5570	ATA5558	T5557
Operating frequency	100 kHz – 150 kHz	100 kHz – 150 kHz	100 kHz – 250 kHz	100 kHz – 150 kHz
Supply current read mode	4 μA	4 μA	7 μA	4 μA
Supply current programming	40 μA		40 μA	40 μA
Maximum DC current	20 mA	20 mA	20 mA	20 mA
Maximum AC current	20 mA	20 mA	20 mA	20 mA
Coil voltage normal mode	3.6 V		4 V	4 V
Coil voltage Read / Write	> 6 V	> 10 V	> 6 V	> 6 V
Coil voltage Programming	> 8 V		> 6 V	> 8 v
Clamp voltage	17 V – 23 V	17 V – 23 V	7 V – 16 V	17 V – 23 V
Internal capacitance	78 pF	340 pF	78 pF	78 pF
Start-up time	< 1 msec	< 3 msec	< 3 msec	< 3 msec
Programming cycles	100,000 cycles	100,000 cycles	100,000 cycles	100,000 cycles
Data Retention	50 years	50 years	50 years	50 years
Maximum storage temperature	-40°C to +150°C	-40°C to +150°C	-40°C to +150°C	-40°C to +150°C
Maximum operating temperature	-40°C to +85°C	-25°C to +105°C	-40°C to +85°C	-40°C to +85°C
Maximum dissipated power	100 mW	100 mW	100 mW	100 mW

3.6 Mechanical Considerations and Transponder Assembly

The integrated circuits or chips in which the transponders are based are available in different packages. While commercial transponders use the chip-on-board (COB) technique described in Chapter 2, this package is not adequate for the development of prototypes. For this reason, most of the transponders are also available in the more common plastic dual-in-line packages (PDIPs) or small outline plastic packages (SOICs).

For example, some of the transponders described in this chapter, such as the MCRF200/250, the MCRF355/360, and the family of MCRF450 transponders, all manufactured by Microchip Technology, are available in these three packages (COB, SOIC, and PDIP). In their SOICs or PDIPs, these are available in eight pins; four of the pins are for testing purposes, two of the pins are for

TABLE 3.6

Summary of Electrical Parameters for Several RFID Transponders Manufactured by Atmel®

Parameter	TK5530	TK5551
Resonance frequency	121.4 kHz – 129.2 kHz	120 kHz – 130 kHz
Inductance	3.95 mH	3.8 mH
Quality factor	13	13
Max field strength before modulation	2 A/m	4 A/m
Field strength operation	30 A/m (-40°C) – 17 A/m (+85°C)	30 A/m (-40°C) – 17 A/m (+85°C)
Field strength programming mode		50 A/m
Modulation Range	4.0 V (20 A/m) – 8.0 V (100 A/m)	4.0 V (20 A/m) – 8.0 V (100 A/m)
Programming time		16 ms per block
Data retention		10 years
Programming cycles		100,000 cycles
Absolute maximum field strength	1000 A/m	1000 A/m
Maximum storage temperature	-40°C to +125°C	-40°C to +125°C
Maximum operating temperature	-40°C to +85°C	-40°C to +85°C

connecting the transponder antenna, and the two remaining pins are not-connect pins. The devices ATA5570 and ATA5567/5558 from Atmel® are also available in COB packages and PDIPs, also using eight pins. The ATA5570 uses four not-connect pins, two pins for connecting the antenna, one pin for ground, and one pin for connecting the output of an external sensor. Because the ATA5567 and ATA5558 do not use an external sensor, their pinout is much simpler using six not-connect pins and two pins for connecting the external antenna. The dimensions of the COB version for all these transponders are 8 mm × 5 mm with a thickness of 0.4 mm. This extremely reduced size makes it ideal for incorporating the transponder and its antenna into the inlay. Transponders, especially in their COB form, are extremely sensitive to issues such as electrostatic discharge and their exposure to ultraviolet light that may result in erasing the contents in their memory cells. However, the use of X-rays for die inspection does not harm the device or erase their memory cell contents.

The process for assembling the whole transponder varies depending on the frequency range at which it will operate as well as the specifications for the final product. For transponders operating in the LF region, the assembly process starts by preparing the die and the capacitor and testing the functionality of the combined die-capacitor element. This is followed with preparing the antenna coil used in this frequency range and creating the inlay formed by the coil and the COB. After the inlay has successfully passed the test, the finished transponder is processed depending on the

TABLE 3.7

Summary of Electrical Parameters for Several RFID Transponders Manufactured by Texas Instruments (Tag-it™)

Parameter	RI-I11-112A	RI-I02-112A	RI-I03-112A	RI-I15-112B	RI-I16-112A	RI-I17-112A
Resonance frequency	13.86 ± 0.2 MHz	13.86 ± 0.2 MH	13.86 ± 0.2 MH	14.1 ± 0.2 MH	13.70 ± 0.4 MH	13.80 ± 0.4 MH
Typical required field for read	96 dBµA/m	94 dBµA/m	107 dBµA/m	98 dBµA/m	113 dBµA/m	110 dBµA/m
Maximum required field for read	101 dBµA/m	97 dBµA/m	109 dBµA/m	101 dBµA/m	116 dBµA/m	113 dBµA/m
Typical required field for write	101 dBµA/m	97 dBµA/m	111 dBµA/m	101 dBµA/m	116 dBµA/m	113 dBµA/m

desired final product, such as card lamination and plastic molding. The last step in this process is a final functional test.

Transponders operating in the HF region can be assembled using three different methods. The first method, which results in the transponder with the COB, is based on the same assembly process as described earlier for LF transponders. A second method results in creating the transponder with direct wire bonding to the antenna that can be etched, printed, or stamped. This assembly method starts by preparing the die; preparing the antenna on the substrate etched, printed, or stamped; and then making the inlay by direct wire bonding to the antenna. After the inlay test, the finished transponder is created by using card lamination, plastic molding, or another method. The last step is a final functional test. The third method creates a transponder with a flip-chip process using a bumped die. The process starts by die bumping for flip-chip assembly, preparing the antenna on the substrate by etching, printing, or stamping it. The inlay is then created using the flip-chip process to attach the antenna to the die. This is also followed by the inlay test, and the finishing of the transponder by lamination or plastic molding. As always, the last step is a final functional test.

4

Antennas for Interrogators

CONTENTS

This chapter describes the principles of design for interrogator antennas. The basic theory that describes the behavior of the antennas used in interrogators is the same that was used to describe the behavior of antennas used in transponders. However, in the case of antennas for interrogators, the size of the antenna is much less critical: because the interrogator has less size restrictions, the antenna can have larger dimensions. This results in an increased antenna performance. Furthermore, antennas for interrogators can be created with thicker conductors that can substation higher currents, resulting in stronger electromagnetic fields. The chapter starts by describing the basic principles used in the design of antennas for interrogators operating in the high-frequency (HF) and low-frequency (LF) bands. It continues by describing auxiliary elements such as tuning indicators, multiplexers, and preamplifiers used with those antennas in order to improve the overall performance of the system. The next section describes the basic principles used in the design of antennas in interrogators operating in the ultra-high-frequency (UHF) band. The last section in this chapter shows different examples of commercially available antennas for interrogators.

4.1 Antennas for HF and LF Interrogators

4.1.1 Design of Antennas for HF and LF Interrogators

Antennas for interrogators operating in the HF and LF ranges are designed for parameters such as maximum read distance for transponders, transponder read rate, as well as regulations such as the maximum field allowed in some specific conditions. The majority of antennas used in HF and LF interrogators are variations of the coil antenna that generates a magnetic field. These antennas are typically called *portal antennas* because due to their physical dimensions and geometry, they can be placed in portals. A small number of antennas for HF and LF interrogators are based on a small dipole that generates an electric field. These antennas are normally called *stick antennas.*

In general, the distance at which a transponder can be read increases with the size of the antenna. However, this size cannot increase arbitrarily. This is due to the practical limitations in using an antenna of a large size and also to the limitations due to regulatory issues. Examples of these factors are the decrease in the signal-to-noise ratio as the size of the antenna increases; the potential for exceeding the regulatory limits of field strength; the need for shielding nearby electronic equipment to avoid unwanted interaction with the generated field; the increase in the area of null zones in which the generated field is zero; and the increase in the inductance of the antenna to a point that makes matching the antenna to the reader extremely difficult, if not impossible.

Interrogators working in the HF range expect an antenna tuned at the frequency of 13.56 MHz, with an input impedance of 50 Ω and a loaded quality factor (Q factor) of less than 20. Similarly, interrogators operating in the LF range expect an antenna tuned to the appropriate LF frequency, an input impedance of 50 Ω, and a loaded Q factor of less than 20. In both cases, for best performance, the match of the interrogator with the antenna should have a *voltage standing wave ratio* (VSWR) lower than 1.2.

The VSWR is a measure of how much of the energy sent through a transmission line is reflected back to the transmitter instead of being transferred to the antenna in order to be converted to the electromagnetic field. Standing waves occur as a result of the existence of incident and reflected waves. The ratio of reflected voltage to incident voltage is called the *reflection coefficient* and is denoted by the symbol Γ. Mathematically, the reflection coefficient is

$$\Gamma = \frac{E_r}{E_i} \tag{4.1}$$

where

Γ = the reflection coefficient

E_r = the reflected voltage

E_i = the incident voltage

The reflection coefficient can also be expressed in terms of the impedance of the transmission line and the impedance of the load as

$$\Gamma = \frac{Z_L - Z_o}{Z_L + Z_o} \qquad (4.2)$$

where Z_L = the impedance of the load and Z_o = the characteristic impedance of the transmission line.

Equation (4.2) shows that when the transmission line is terminated with a load with an impedance equal to its characteristic impedance, the reflection coefficient is zero. This means that all the energy present in the line has been transferred to the load. When the line is terminated with either an open circuit or a short circuit, the absolute value of the reflection coefficient is equal to 1. This indicates that none of the energy present in the line has been transferred to the load, but is being reflected back to the transmitter. For any other terminations, including complex impedance values, the value of the reflection coefficient is, in absolute value, between 0 and 1. This indicates that a portion of the energy present in the line has been transferred to the load and some other portion of the energy remains in the line.

The VSWR is defined as the ratio of the maximum voltage in the line (E_{max}) compared to the minimum voltage in the line (E_{min}):

$$VSWR = \frac{E_{max}}{E_{min}} \qquad (4.3)$$

The VSWR can also be defined as a function of the reflection coefficient (Γ):

$$VSWR = \frac{1+|\Gamma|}{1-|\Gamma|} \qquad (4.4)$$

Equation (4.4) shows that in the case of maximum reflection due to the transmission line terminated in an open or a short circuit, which results in $|\Gamma| = 1$, the value of the VSWR is infinity. Conversely, when the line is terminated with its characteristic impedance that results in $|\Gamma| = 0$, the value of VSWR is equal to zero.

Example 4.1:

The designer of a RFID system specifies a match between an interrogator and the antenna, resulting in a VSWR better than 1.2 using a 50 Ω coaxial cable as the transmission line. Calculate the maximum reflection coefficient and the allowed range of antenna impedance, assuming it is perfectly resistive.

Solution:

The relationship between VSWR and the reflection coefficient is given by equation (4.4). Algebraic manipulation of this equation yields $|\,\Gamma\,| = 0.1$. This is the maximum value allowed for the reflection coefficient.

The relationship between the reflection coefficient and the impedance of the load is given by equation (4.2). When solving for the value of Z_L using this equation, it is necessary to consider the two signs associated with the previous result of $|\,\Gamma\,| = 0.1$ as the impedance of the load can exceed or be below the impedance of the line.

$\Gamma = 0.1$ results in $Z_L = 61\ \Omega$, while $\Gamma = -0.1$ results in $Z_L = 41\ \Omega$. Therefore, an antenna impedance ranging from 41 Ω to 61 Ω will result in a VSWR better than 1.2.

Note than the range of allowed load impedances is not symmetric referred to the impedance of the line (Z_o). There is more tolerance for values above Z_o than for values below Z_o.

A transmission line that is not perfectly matched (i.e., a VSWR value higher than 1) reduces the amount of power from the interrogator that reaches the antenna. More importantly, however, is the fact that a fraction of the power that remains in the transmission line can break down the dielectric of the line or increase its temperature due to the power dissipated. Therefore, it is critical to achieve the best possible matching.

Antenna analyzers are a very versatile and useful tool at the time of designing antennas and matching them to their transmission lines. Figure 4.1 shows an image of a low-cost antenna analyzer. The configuration of this analyzer is done through software, and the results are displayed and stored in a personal computer.

FIGURE 4.1
Antenna analyzer from Array Solutions. The control of the unit and the display of results are done by software.

This unit shown in Figure 4.1 allows the measurement of the complex impedance of a transmission line or an antenna, the measurement of reflection coefficient, insertion losses, and other parameters of interest. The front panel only contains an ON/OFF switch, two light-emitting diodes (LEDs) that indicate whether the unit is energized and whether the unit is running a test, and a single Bayonet Neill-Concelman (BNC) connector. The back panel contains the power supply connection and a DB9 connector for a serial cable from the control computer to configure the instrument and transmit the results. The user can select the range of frequencies for the sweep-starting frequency, ending frequency, and step- or central frequency and bandwidth. This unit can also act as a fixed radio frequency source by generating a user-selected single frequency signal. The output values are shown graphically on the screen as conventional graphs (parameter versus frequency) or Smith charts. The data can also be stored as a text file for further processing. The graph can display one or more of the following parameters: impedance magnitude, impedance phase, reflection coefficient, return loss, stationary wave ratio, series load circuit, or parallel load circuit. Figure 4.2 shows an example of the output showing the magnitude and phase of a transmission line between 25 MHz and 40 MHz.

Figure 4.2 depicts the magnitude and phase of the impedance measured by the antenna analyzer for a transmission line with an electrical length

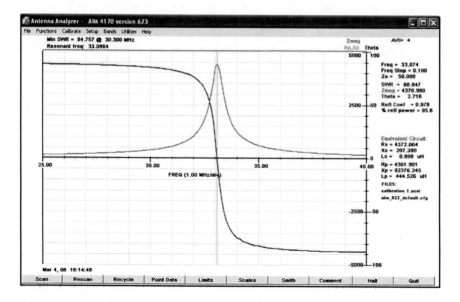

FIGURE 4.2
Frequency behavior of a transmission line terminated with a short circuit for frequencies around a quarter wavelength. This display shows the magnitude and the phase of the impedance at the near end.

equal to one-quarter of the wavelength, terminated in a short circuit. Due to the impedance transformations at this specific electrical length, the theoretical value of the transformed impedance seen at the near end of the line is infinity. In practice, the losses of the line make the impedance have a large, yet finite, value. The output also displays other values described earlier in this section, such as the reflection coefficient and the VSWR, among others. Similar results could also be obtained using other commercially available antenna analyzers.

When designing the antenna, it is also necessary to take into consideration the write range of transponders that is about 70% of their read range. This is due to the higher amount of power needed to rewrite a transponder compared with just reading it. The orientation of the antenna with respect to the transponder is also a factor in determining the maximum read range in a given application. As shown in Figure 4.3, the field generated by the coil antenna does not have the same strength in all the spatial directions, but instead it has regions of maximum values alternating with regions of minimal strength.

An additional parameter that may limit the maximum read range of a system is the presence of metal objects in the vicinity of the antenna. Metal objects close to the antenna have the effect of detuning it, and therefore it loses its efficiency at the desired frequency. As the size of the antenna increases, the minimum required separation distance between the antenna and the metal objects also increases. For antennas operating in the HF frequency range, the minimum distance between antenna and metal has to be greater than 10 cm.

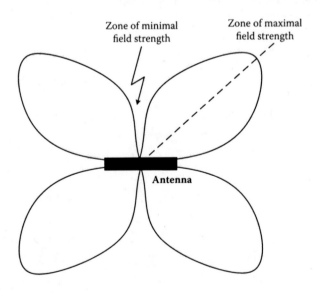

FIGURE 4.3
The radiation diagram for the loop antenna in the interrogator shows areas of maximum and minimum field strength.

Even at 30 cm of separation, there is a considerable decrease in the reading range. When the distance between antenna and metal object reaches 50 cm, there is no measurable difference. Similarly, the presence of other antennas in the vicinity also affects the overall performance of the system.

Antennas can be built using any conductive material. However, they are normally built using copper or aluminum. The type of material used for the antenna and the dimensions of the tube or strip lines depend on the desired mechanical parameters of the antenna as well as the maximum inductance that must be matched to the interrogator. In addition to the conductors, the matching process requires the use of capacitors and resistors. These must be rated for the high voltages that may arise in resonant antennas as well as for the power that they must dissipate. Typical power values for these resistors range from 2 W to 50 W.

The dimensions of the antenna have an effect on the read distance between interrogators and transponders. The strength of the magnetic field for an electrically small-loop antenna is

$$B_z = \frac{\mu_0 I N a^2}{2(a^2 + r^2)^{3/2}} \quad (Weber \, / \, m^2) \tag{4.5}$$

with a being the radius of the loop and r being the radial distance from the center of the loop. To study the effects of the antenna radius, the rest of the parameters in equation (4.5) can be considered constant. Assuming that all antennas have the same amount of current and also the same number of turns, the only parameter that changes is the radius of the loop. Figure 4.4 depicts the values of magnetic fields for coils with three different radii (0.2 m, 0.4 m, and 0.8 m). Because the graph is only intended to highlight the dependence of the field with the distance from the center of the coil, it uses arbitrary units for the strength of the field depicted in the vertical axis.

It is interesting to note that, although larger coils produce stronger fields, this only happens after a certain distance threshold has been reached. In particular, for read distances closer to the antenna, the stronger fields are produced by the coils with the smallest radius. Antennas with smaller radii produce stronger fields in the vicinity of the antenna, but these fields decrease faster for longer distances. Therefore, the design of the antenna must consider the range of distances between transponders and the interrogator when selecting the size of the loop.

The resonant frequency of an antenna (ω_o or f_o) is

$$\omega_o = \sqrt{\frac{1}{LC}} \quad or$$

$$f_o = \frac{1}{2\pi\sqrt{LC}} \tag{4.6}$$

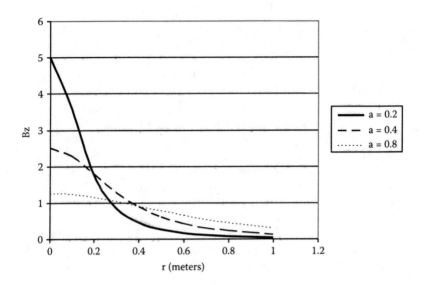

FIGURE 4.4
Strength of the magnetic field as a function of the distance from the coil for loop antennas of different diameters.

An antenna will reach its resonant frequency when its inductive impedance is equal to its capacitive impedance. The loop antenna has an inherently inductive behavior. Therefore, it needs an external capacitor to provide the capacitive impedance required for resonance. The value of this external capacitor can be calculated using equation (4.6). It is important to note that the capacitance and inductance of the antenna are inversely related. A problem may arise if the inductance of the antenna increases to a point that may make it difficult to use a capacitor of suitable value. For example, a loop antenna with an inductance of 5 µH operating in the HF range requires a matching capacitor of 27 pF. For these small capacitance values, the parasitic capacitances that have not been taken into account when modeling the antenna may contribute significantly to the overall capacitance and therefore modify its resonant frequency. It is then important for antennas to be designed with lower inductive values. This may require one to use, for example, low-resistance conductive tube instead of wire at the expense of increasing its physical dimensions.

The inductance of the antenna made with copper tube, shown in Figure 4.5, can be estimated as

$$L(\mu H) = 0.008\ a\left[\ln\left(0.707\ \frac{a}{d}\right) + 0.379\right] \tag{4.7}$$

where
  L = the estimated value of inductance
  d = the diameter of the tube in cm

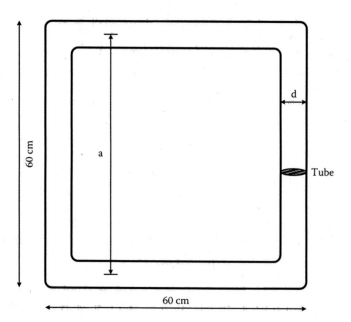

FIGURE 4.5
Basic structure of a loop antenna made with copper tube.

    a = the length of the side of the antenna from center to center (as shown in Figure 4.5)

Example 4.2:

An interrogator operating at HF uses a loop antenna made of copper tube with a diameter of 2 cm, with a rectangular shape of 60 cm × 60 cm. Estimate the value of the inductance for the antenna. Estimate also the value of the capacitor required to tune the antenna for resonance.

Solution:

Using equation (4.7) with d = 2 and a = 59 yields L = 1.6 μH. This value of inductance can now be used in equation (4.6) to calculate the estimated value of capacitance. The estimated value of capacitance is 86 pF.

    The previous equations provide an estimated value of antenna inductance. After the antenna has been built, its inductance must be measured in order to choose the correct value for the tuning capacitor. This can be done with an impedance analyzer that allows the selection of the measurement frequency. However, impedance analyzers are expensive and may not be available in most cases. In this case, a low-cost LCR (inductance [L], capacitance [C], and resistance [R]) meter that operates at the frequency of 1 kHz can give a reasonable approximation of the antenna inductance. The antenna analyzer shown in Figure 4.1 is another low-cost alternative for measuring the inductance of the antenna and the resonant frequency of the antenna-capacitor system.

The last parameter that needs to be calculated and measured when designing the antenna is its Q factor. The Q factor is a measure the frequency selectivity of an antenna. The Q factor is related to the range of frequencies above and below its resonant frequency that the antenna will be able to transmit or receive with losses below 3 dB. The Q factor can be estimated as

$$Q = \frac{f_r}{BW_{-3dB}} \qquad (4.8)$$

where Q is the quality factor of the antenna, f_r is its resonant frequency, and BW_{-3dB} is the bandwidth of the antenna at 3dB of attenuation.

A high Q factor indicates an antenna that is very narrow in frequency, that is, highly selective. This may be good in order to reject interfering signals that otherwise would be detected by the antenna. However, the RFID signal has a certain bandwidth. If the antenna is too selective, it may distort the components of the transmitted signal that fall outside the selectivity of the antenna, as shown in Figure 4.6. A good experimental compromise is to choose a Q factor of around 20.

If the Q factor of the system made of the antenna and capacitor is higher than 20, it needs to be brought down by using a dampening resistor in parallel with the antenna in order to widen its bandwidth. This technique is shown in Figure 4.7.

The introduction of the dampening resistor produces Ohmic losses in the system, lowering the Q factor. The Q factor is related to the parallel resistance as

$$Q = \frac{R_{par}}{2\pi f L} \qquad (4.9)$$

An additional problem arises when using equation (4.9) in order to select the value of the parallel dampening resistor: the value R_{par} in the equation is the combination of the external dampening resistor (R_{damp}) in parallel with the resistance of the antenna at the frequency of interest (R_{ant}). The measurement of the resistance of the antenna at the operating frequency is difficult and requires the use of expensive instrumentation. However, the Q factor for an antenna is relatively easy to measure experimentally with an antenna analyzer or using a radiofrequency signal generator and a spectrum analyzer. Therefore, the resistance of the antenna R_{ant} at the operating frequency can be calculated indirectly by measuring the Q factor of the antenna without an external dampening resistor. Once this value has been found, it is now possible to calculate the required value of the total parallel resistance R_{par} to achieve the desired value of Q. Because R_{par} is the parallel of R_{ant} and the

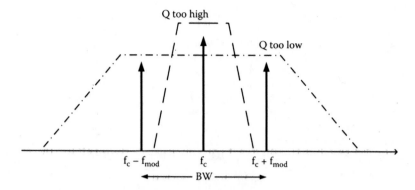

FIGURE 4.6

Q factor of an antenna and its relationship with the antenna performance. (a) A very low quality factor (Q factor) does not provide adequate selectivity. (b) A very high Q factor may reject components of the transmitted signal.

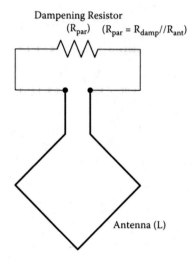

FIGURE 4.7

Dampening resistor connected in parallel with the antenna. This is used to lower the Q factor to the desired value.

external dampening resistor R_{damp}, it is now possible to estimate the value of this external resistor. Example 4.3 illustrates this procedure.

Example 4.3:

Consider the antenna for the RFID system from Example 4.2. Estimate the value of the external dampening resistor necessary for the system to achieve a Q factor of 20. The initial measurement of the Q factor for the antenna-capacitor system resulted in a value of 55.

Solution:

Using equation (4.9) with a Q factor of 55, an inductance value for the antenna of 1.6 μH, and a frequency of 13.56 MHz results in a resistance equal to 7.5 kΩ. Because the Q factor of 55 was measured without the use of external resistors, the value of resistance obtained is the resistance of the antenna: $R_{ant} = 7.5$ kΩ.

Now, we must calculate the value of the total parallel resistance needed to reduce the Q factor to the desired value of Q. We use again equation (4.9) with the same operational parameters as before, with the exception that the new value for Q results in a resistance equal to $R_{par} = 2.7$ kΩ.

Considering that the total value of the resistance found above is equal to the resistance of the antenna in parallel with the dampening resistor ($R_{par} = R_{ant}$ // R_{damp}), it is now possible to calculate the value of the external dampening resistor. This results in $R_{damp} = 4.2$ kΩ. Because this is not a standard resistor value, the designer should use the next standard value, which is 4.3 kΩ.

Therefore, the addition of 4.3 kΩ results in lowering the Q factor from its initial value of 55 to the desired value of 20. Because the value of inductance used for this calculation could be an estimated value, the design process involves measuring the Q factor with the external dampening resistor of 4.3 kΩ and adjusting its value if necessary to reach a Q factor of 20. Finally, it is also necessary to keep in mind that this external dampening resistor must be sized to dissipate the power transmitted by the interrogator.

4.1.2 Antenna Matching

Matching the impedance of the antenna to the impedance of the transmission line (normally a coaxial cable of 50 Ω) ensures that all the power from the interrogator is transferred to the antenna, and there is no power reflected in the line. In practice, a VSWR of 1.2 or lower is acceptable as the majority of the power is transmitted to the antenna. From the several approaches to match an antenna to its transmission line, the most widely used matching methods for interrogator antennas in RFID systems are gamma matching, T-matching, transformer matching, and capacitance matching.

Gamma matching is a technique based on connecting the two conductors of the coaxial cable that make up the transmission line to the two points in the antenna that exhibit a resistive impedance equal to 50 Ω. The shield of the coaxial cable is normally connected to the point in the antenna that is opposite to the gap used to connect the resonance capacitor and the dampening resistor. The central connector of the coaxial cable is then connected, by means of an additional conductor, to the appropriate point in the antenna as shown in Figure 4.8.

In practice, the point at which the central conductor of the coaxial cable is connected (point A) must be determined by experimentation. The designer of the system connects an antenna analyzer at the far end of the coaxial cable and moves point A until the antenna analyzer reads a VSWR value equal to 1. This indicates that the impedance of the antenna between point A and the connection of the central conductor is equal to 50 Ω. The value of the capacitor used to tune the antenna to the resonance frequency was

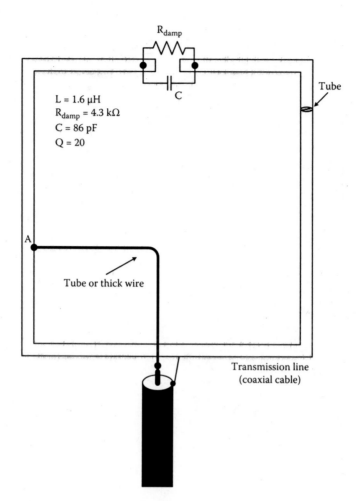

FIGURE 4.8
Connections for gamma matching.

determined using an approximation for the value of the antenna inductance. Therefore, a variable capacitor is initially used in order to tune the antenna to its resonant frequency. Once the correct value of capacitance has been determined, it is then possible to substitute the bulkier variable capacitor with a fixed capacitor.

T-matching is similar to gamma matching with the difference being that in T-matching, both the internal and external conductors of the coaxial cable are tapped to two points in the antenna, as shown in Figure 4.9. It is based on the fact that two points of the antenna, equidistant from the center, exhibit a purely resistive impedance.

The two tapping points are located symmetrically with respect to the gap used for connecting the tuning capacitor and the dampening resistor. The

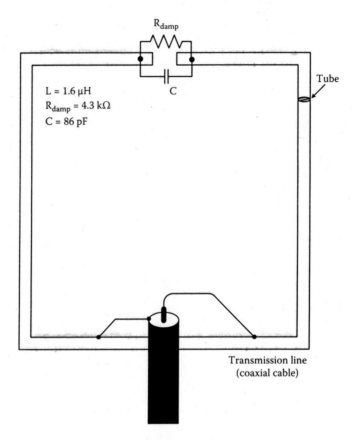

R_{damp}

C

Tube

L = 1.6 µH
R_{damp} = 4.3 kΩ
C = 86 pF

Transmission line
(coaxial cable)

FIGURE 4.9
T-matching.

two tapping points are found by using an antenna analyzer until the VSWR value is equal to 1. Higher values of the Q factor results in the two matching points being closer together.

Transformer matching has the advantage that it electrically isolates the antenna from the interrogator, thus breaking the Ohmic continuity between these two parts of the system. This is useful when the RFID system is affected by noise as it breaks any ground loop that may exist. Figure 4.10 shows the electric diagram of transformer matching. It is important to note that the matching occurs at the gap in the antenna that is used to connect the resonance capacitor and the damping resistor instead of their opposite point, as it happens with gamma and T-matching.

Transformer matching requires the design of two elements: the matching transformer and a balum (balanced–unbalanced) transformer. The design of the matching transformer requires one to size the number of turns of the primary (N) that is connected to the transmission line and the secondary (M)

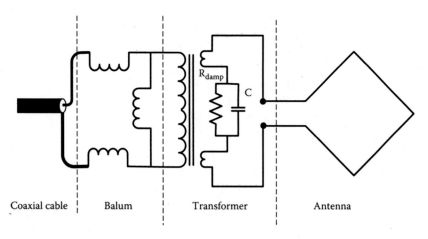

FIGURE 4.10
Transformer matching. Notice how the matching point is the same point at which the capacitor
and resistor are connected.

that is connected to the antenna. The ratio of turns can be found using the
following:

$$\left(\frac{N}{M}\right)^2 = \frac{Z_o R_{par}}{(2\pi f L)^2} \tag{4.10}$$

where (N/M) is the primary to secondary ratio of turns, Z_o is the impedance
of the transmission line, R_{par} is the parallel resistance of the antenna and the
dampening resistor, and L is the inductance of the antenna.

Example 4.4:

Using the values from the previous examples in this chapter, calculate the number
of turns for the primary and secondary of the matching transformer. Assume a
characteristic impedance of 50 Ω.

Solution:

Using equation (4.10) with R_{par} = 2.7 kΩ, L = 1.6 μH, and f = 13.56 MHz, the ratio
of turns is equal to 2.7. Using, for example, M = 3 results in N = 8 windings.

The balum transformer is used to connect unbalanced transmission lines
such as a coaxial cable to balanced antennas such as the loop antenna.
Although they can be connected directly, as with the gamma matching,
the common mode currents that appear in the antenna and line due to
the unbalance may cause excessive noise and disturb the circuits. In this
case, the connection of a balum transformer between line and transformer

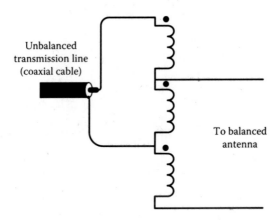

FIGURE 4.11
A balum is used to connect unbalanced transmission lines to balanced antennas.

solves these problems. The structure of a balum transformer is shown in Figure 4.11.

Both the matching transformer and the balum transformer are normally made with a toroid ferrite core. The balum transformer can be tested by connecting a 50 Ω load to the balanced side and measuring the VSWR through a coaxial cable connected to the unbalanced side. The VSWR should be close to 1.0.

Capacitance matching attempts to match the antenna and its transmission line by adding to external capacitors between the resonance capacitor and the dampening resistor, as shown in Figure 4.12.

Because of the stray capacitances inherent to the system, it is very difficult to predict the adequate values for the two matching capacitors, which requires extensive experimentation. Furthermore, for larger antennas, the values of the two matching capacitors tend to be very small, making the matching process very sensitive to stray capacitances. In Figure 4.12, the series combination of C_1 and C_2 must be equal to the capacitance value required to tune the antenna at the correct frequency.

The value of the matching capacitor C_{m1} can be estimated as

$$C_2 = C_{tuning} \sqrt{\frac{R_{par}}{R_{out}}} \tag{4.11}$$

where C is the total capacitance required for tuning the antenna, R_{par} is the equivalent value of the AC resistance of the antenna in parallel with the damping resistor, and R_{out} is the desired matching impedance, typically 50 Ω. The value of C_1 can be calculated through the series capacitance of C_1 and C_2.

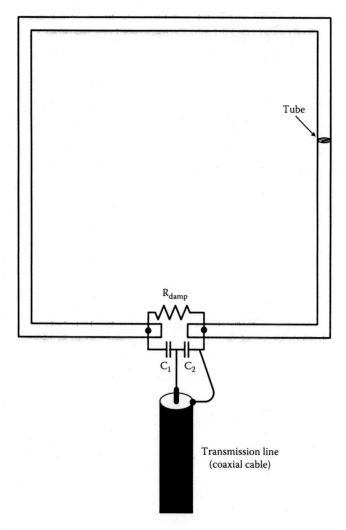

FIGURE 4.12
Capacitance matching.

Example 4.5:

The antenna used in Example 4.4 must be matched to a 50 Ω transmission line using capacitance matching. Calculate the values of the two matching capacitors required.

Solution:

The required tuning capacitor has a value of 86 pF, and the parallel resistance has a value of 2.7 kΩ. The required value for R_{out} is 50 Ω. Using equation (4.11) yields a value of $C_2 = 632$ pF. The equivalent series capacitance is 86 pF, allowing one to calculate the value of C_1 equal to 99 pF.

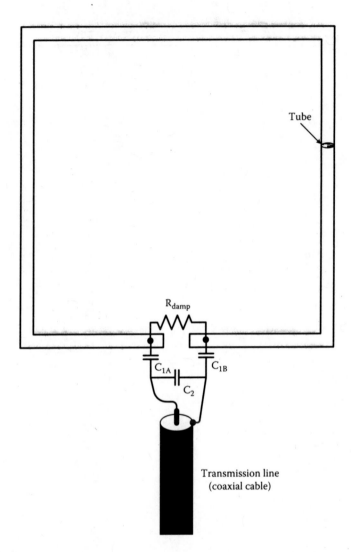

FIGURE 4.13
Balanced capacitance matching.

Some antennas require a matched capacitance so the capacitance seen by both conductors of the coaxial cable is the same. This can be achieved by using the structure shown in Figure 4.13. In this structure, the value of C_1 is doubled and added to each side of C_2.

Example 4.6:

Using the results of Example 4.5, calculate the capacitor values required for balanced matched capacitance.

Solution:

From Example 4.5: $C_1 = 99$ pF and $C_2 = 632$ pF. To achieve balanced capacitance, $C_{1A} = C_{1B} = 198$ pF. C_2 remains at the same value of 632 pF.

4.1.3 Multiple Antenna Systems

It is not uncommon for several antennas to be located in close proximity. This may be due to different interrogators being close to each other or for multiple antennas to be connected to a single interrogator. When designed correctly, the mutual coupling between the antennas can enhance the performance of the overall system. However, a careless positioning of the multiple antennas may result in their mutual coupling, severely degrading their performance.

When several single-antenna interrogators coexist in close vicinity to each other, it is necessary to minimize their interaction with each other. The coupling between these antennas depends on the distance between them as well as their relative angles. Although the radiation pattern of a single antenna is well known and easy to predict, this radiation pattern becomes distorted due to the coupling. The best approach for locating the optimal placement of two or more antennas is by using an antenna analyzer in order to find the minimum of radiation by measuring the voltage generated in the nearby antenna.

A technique sometimes used to enhance the radiation pattern of the antenna in a single interrogator is to place another matched but unconnected antenna, opposite to the antenna connected to the interrogator. This connected antenna is a *driven antenna*, while the unconnected antenna is a *reflective antenna*. The combination of both antennas directs the energy from the driven antenna to a narrower area. Once again, an antenna analyzer should be used in order to determine the best arrangement between the two antennas.

Instead of leaving the second antenna unconnected, making it a reflective antenna, it is also possible to drive it with the signal from the interrogator by means of a splitter. In this case, the output of the interrogator is connected to the input of the splitter, and each output from the splitter is connected to a matched antenna. Splitters also offer the possibility to have the two output signals in-phase or out-of-phase. Splitters introduce losses, but when the two signals are connected in-phase, the reading distance between interrogator and transponders increases more than the reading distance with just one antenna without the losses from the splitter. The increase in the reading distance is at the cost of reducing the radiation pattern in the directions perpendicular to the arrangement of antennas. Therefore, this technique is only useful when the transponders are located only in known areas. When the signals fed to the two antennas are out-of-phase, the radiation pattern changes significantly, resulting in a pattern opposed to when the two signals are in phase. This gives the designer of the system the ability to configure the radiation pattern depending on the needs of the system.

Other techniques based on systems with multiple antennas are directed toward specific applications. In one of them, two identical antennas located in a crossed arrangement are connected to the output of a splitter but use different lengths of transmission line. In particular, if the different between lengths is a quarter of a wavelength, the two fields are shifted in 90°. This produces a field that changes the direction of its maximum, giving the appearance of a rotating field. Some other commercial antennas have an additional input that allows connecting an out-of-phase signal from a complementary antenna in order to maximize field strength in a single direction.

4.2 Auxiliary Elements for LF and HF Interrogator Antennas

4.2.1 Antenna-Tuning Indicators

Tuning indicators are used to facilitate the process of antenna tuning without the need to use antenna analyzers or other instrumentation. They present the advantage of being a self-contained system that has been optimized for specific interrogators and antennas. On the other hand, because of their specific nature, they can only be used with a limited number of antennas.

For example, Texas Instruments manufactures the RI-ACC-ATI2 antenna-tuning indicator used to help tune their LF antennas. The tuning indicator is connected to the interrogator and indicates to the user if the inductance or capacitance in the interrogator needs to be increased or decreased until the optimal value has been achieved. Some interrogators have a tuning capacitor, while some others have a tuning inductance that can be adjusted for optimal performance.

4.2.2 Antenna Multiplexers

Antenna multiplexers are used to connect multiple antennas to a single interrogator. This can be done to take advantage of the multiple coupling between antennas or when it is necessary to read data from multiple interrogators located at distant points not covered by a single antenna. The majority of multiplexers in the market work with two, four, or eight antennas. The multiplexer is connected between the radiofrequency output of the interrogator and the antennas, as shown in Figure 4.14.

It is not necessary for all the antennas connected to the multiplexer to have the same inductance. This presents the advantage of being able to choose the antenna most adequate for a specific application without having to physically connect and disconnect them from the interrogator. However, because most multiplexers do not include a tuning feature, each antenna must be tuned to the appropriate resonance frequency with their own individual tuning method described in Sections 2.2 and 2.3.

Antenna multiplexers are normally created by the manufacturers of interrogators, although there are some third-party companies that also

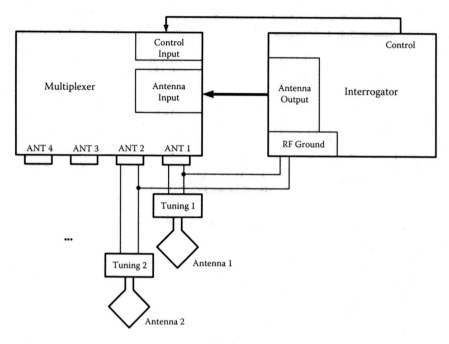

FIGURE 4.14
Multiplexer for connecting multiple antennas to a single interrogator.

manufacture them. For example, Skyetek manufactures multiplexers for the HF frequency range that can control four or eight antennas, having insertion losses lower than 0.7 dB and isolation between adjacent channels better than 45 dB.

4.2.3 Preamplifiers

Increasing the reading distance between interrogators and transponders requires a double approach. First, the electromagnetic field transmitted by the interrogator has to increase in order to generate enough power to turn on the distant transponders. This is relatively easy to do and is mainly limited by standards and regulations on maximum power and maximum allowed field strength. Second, reading distant transponders requires the interrogator to detect signals closer to the noise level. The increase in the detection sensitivity can be achieved with the use of preamplifiers, as shown in Figure 4.15.

The factor that limits the read range in RFID is ultimately the inherent noise that accompanies the signal received by the interrogator. A correctly designed preamplifier is able to increase the level of the signal without degrading the signal-to-noise ratio. The design of the main preamplifier is normally based on a low pass filter in order to reject the frequency generated by the field and

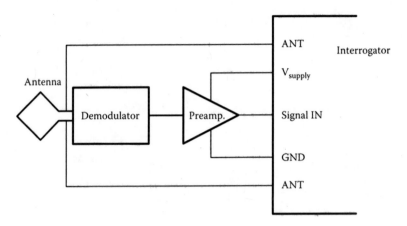

FIGURE 4.15
Preamplifier for extending reading distance.

allow the transmission of data. This stage is followed by an AC amplifier that increases the signal levels to the point that can be easily understood by the interrogator. The design of the preamplifier must consider the trade-offs between increasing the reading distance and the settling time required by the system. This means that the interrogator will be optimized for reading distance or for settling time, but cannot be optimized for both parameters.

Even when using preamplifiers, it is necessary to minimize the other potential sources of noise in the system. These can be the use of bad connectors between the interrogator and the antenna, improper installation of the transmission line between interrogator and antenna, an antenna that is not tuned correctly, the existence of interfering signals from nearby electronic equipment, and noise emanating from the power supply lines, among others. The designer of the RFID system should ensure that all these sources are minimized or at least known, and their values measured.

4.3 Antennas for UHF Interrogators

4.3.1 Design of Antennas for UHF Interrogators

Given the wavelength of signals in the UHF range, it is possible to design antennas for interrogators that operate in the far-field region as interrogators do not have the same space limitations of transponders. The resulting radiation pattern of two dipoles located at a short distance from each other can be easily modified by carefully arranging the distance between them as well as the phase shift of the signals that they radiate. A further modification of the radiation pattern comes from placing a large metal surface at a specific distance from the two dipoles. The metal surface acts like a shield that effectively removes half of the radiation pattern. Furthermore, if the distance between the antennas and the metal shield is one-fourth of the signal wavelength, the phase shift due to the distance between antenna and shield, combined with the phase shift created by the reflection in the metal shield, places the reflected signals in phase with the signals originated by the antenna. This, in turn, has the effect of doubling the power transmitted in the area away from the metal shield, as shown in Figure 4.16. It is necessary to remember that the increase of power in that specific direction comes at the expense of having no power in the area behind the metal shield, also known as a *passive reflector*. This structure is commonly used in *patch* or *flat* antennas.

An additional advantage of working in the far-field region is the possibility to take advantage of polarization. Polarized signals have their electric fields oscillating in a known and constant direction in space. This way, the signal recovered by the receiving antenna has a marked dependence on the geometrical relationship between the direction of the electric field and the direction of the antenna. Therefore, the energy recovered by the antenna in the interrogator ranges from a maximal value to zero depending on the relative orientation between the antennas in the interrogator and transponder. Using polarized signals allows discriminating between transponders. When the discrimination between transponders is

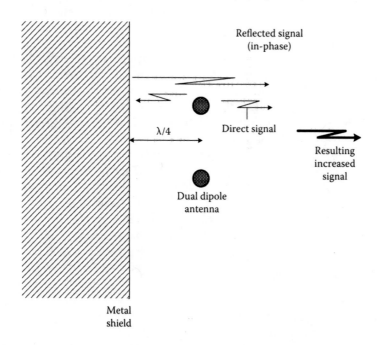

FIGURE 4.16
Metal shield increases effective radiated power in one specific direction.

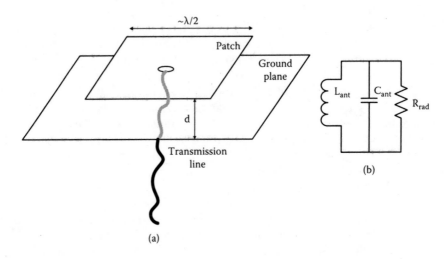

FIGURE 4.17
(a) Basic structure of a patch antenna. (b) Parallel equivalent circuit for a patch antenna.

not required, the use of dual dipole antennas in the transponders ensures that the interrogator will receive an adequate amount of energy.

The design of an antenna for the UHF RFID band represents an increased engineering challenge compared to designing it for the HF or LF bands because of the higher bandwidth needed for signals in the UHF band. It also becomes more difficult to maximize the performance of the antenna over the wider frequency range allocated for RFID systems in the UHF band. For this reason, while most of the HF and LF antennas follow the same basic design, engineers have developed several approaches to the design of UHF antennas for interrogators based on different performance parameters. This is especially critical for handheld applications in which the limited size of the antenna degrades its performance.

The patch antenna shown in Figure 4.17(a) is one of the most commonly used antennas for UHF interrogators. In its basic form, it consists of a metal patch of approximately one-half of a wavelength long and mounted over a ground plane. Figure 4.17(b) shows its equivalent electrical model.

Typical values for the equivalent circuit shown in Figure 4.17(b) are around 150 Ω for the radiation resistance of the antenna, between 9 and 10 pF for its capacitance, and from 2 to 4 nH for its inductance. The matching of the transmission line with a 50 Ω characteristic impedance to the relative high radiation resistance of the antenna is done by changing the location at which the transmission line feeds into the metal patch away from its center. The bandwidth of the patch antenna is a function of its dimensions and the separation between the metal patch and the ground plane. The type of dielectric used between the planes is also important, as dielectrics other than air will increase its equivalent capacitance.

The patch antenna, by its very own design of using a ground reference as part of the antenna, is an unbalanced antenna. Because unbalanced antennas do not operate well when they are close to dielectric materials, and the human body can be essentially considered as a dielectric material at UHF frequencies, patch antennas are not the best choice for handheld interrogators. Furthermore, the size and weight limitations required by mobile and handheld interrogators further limit the use of patch antennas in these applications. A better solution for handheld and mobile interrogators are balanced antennas, that is, antennas with symmetric and opposed currents as shown in Figure 4.18.

When the antenna is fed using an unbalanced transmission line, such as a coaxial cable, it is necessary to use a balum transformer to match the unbalanced line to the balanced antenna. The basic theory behind this balum is the same as was described for HF and LF antennas. Although the structure shown in Figure 4.18 is technically feasible, an additional problem may arise due to the size required by half of a wavelength, especially for handheld interrogators. At the frequency of 915 MHz, the length of half of a wavelength is equal to 16 cm. While this is a very reasonable distance for a stationary antenna, it may be excessively large for applications involving handheld readers. In this case, it is possible to bend the dipoles in a similar way as is done for the antennas in transponders, as shown in Figure 4.19.

The reduction in size for the bent dipoles comes at the expense of degrading the electrical specifications of the antenna. In particular, its radiation resistance decreases and therefore the bandwidth of the bent dipole structure also decreases compared to the bandwidth of the regular dipole.

A different type of antenna used in UHF interrogators (either stationary or mobile) is based on the Yagi-Uda antenna. This structure, shown in Figure 4.20, consists of a driven element and two or more parasitic elements: one reflector and one or more directors. The parasitic elements direct the majority of the

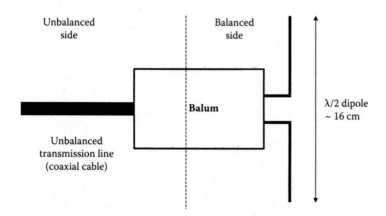

FIGURE 4.18
Balanced half-wavelength antenna.

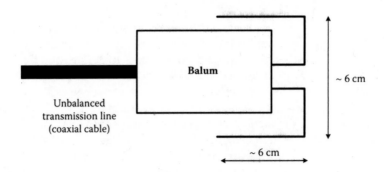

FIGURE 4.19
Bent dipole for a UHF interrogator antenna.

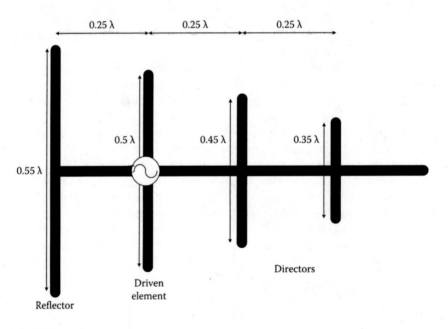

FIGURE 4.20
Yagi-Uda antenna.

energy radiated by the antenna toward the director. Therefore, this structure helps to minimize the interaction of the radiated field with a person who may be holding an antenna for handheld interrogators. Moreover, because this is a balanced antenna, this interaction between the energy radiated by the antenna and that radiated by the human body is further minimized.

In its original form, however, Yagi-Uda antennas can still be considered too large to use with handheld interrogators. In this case, it is possible to bend the dipole and longer elements, as described above. The existence of the parasitic elements will provide additional gain compared to the single dipole.

4.3.2 Transmission Lines and Connectors for UHF Interrogators

Contrary to LF and HF systems, the coaxial cables and connectors typically used in UHF RFID systems have the potential to extremely degrade the signal they transmit. The main reason for this problem is the skin depth. *Skin depth* describes the effect that the current flowing through a conductor flows through a portion of its inner surface. Equation (4.12) shows that the skin depth (δ) is inversely proportional to the frequency of the signal:

$$\delta = \sqrt{\frac{1}{\pi \mu \sigma\, f}} \tag{4.12}$$

where δ is the skin depth, μ the magnetic permeability of the conductor, σ its conductivity, and f the frequency of the current flowing through the conductor.

Therefore, an increase in the signal frequency results in a smaller portion of the conductor being used to transmit the signal. For a typical conductor made of copper, the skin depth for an LF signal is 0.2 mm, the skin depth for an HF signal is 55 µm, and for a UHF signal it is 2 µm. Consider a copper conductor with a radius of 0.2 mm. At LF frequencies and below, the current will flow through all the conductor area. When the frequency increases to 13.56 MHz, only the first 55 µm are actually available as a path for the current. The effective area has been reduced to a half of the area available for an LF signal. Because the conductivity of the conductor depends on its area, the conductivity of this conductor at HF is about half of its conductivity at LF. The situation changes more dramatically at UHF frequencies, as the skin depth of 2 µm means that current can flow only through the outer 1 percent of the conductor. With this, the conductivity of the copper wire for UHF signals has been reduced in a factor of 62 compared to the conductivity at LF. This strong decrease in conductivity introduces considerable losses that may render the system unusable. Figure 4.21 further illustrates this situation.

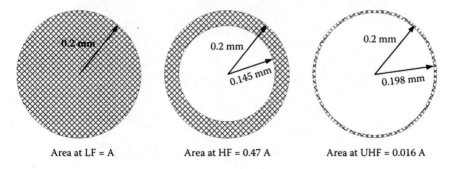

Area at LF = A   Area at HF = 0.47 A   Area at UHF = 0.016 A

FIGURE 4.21
Skin effect at (a) low frequency (LF), (b) high frequency (HF), and (c) ultra-high frequency (UHF). Current flows through the shaded areas.

With these considerations, the layout of the transmission line used to connect the UHF interrogator to its antenna becomes critical. The most effective approach to improving the performance of the system is to minimize the length of transmission line between the output of the interrogator and its antenna. It can also be possible to use transmission lines with larger diameter wires; however, the slight increase in effective area does not compensate for the additional cost of these lines or their increased rigidity. It is also important to choose a coaxial cable that has a continuous outer conductor instead of a multistrand ribbon for its shield.

The use of adequate connectors is also critical for RFID systems working in the UHF band. The BNC connectors that are commonly used in the LF and HF bands could be used without significant performance degradation in UHF applications. However, other types of connectors, specifically designed for higher frequencies, produce more reliable results. TNC connectors were conceived as a threaded version of the BNC connectors, being suitable for use up to 10 GHz. Type-N connectors are commonly used in high-frequency applications as they are suitable to carry signals up to 10 GHz and even 18 GHz with their recent enhancements. They are also mechanically robust and well suited for outdoor use. Figure 4.22 shows a male type-N connector, while Figure 4.23 shows a female type-N connector.

Because type-N connectors are available in 50 Ω and 75 Ω versions and the connectors are typically not labeled, it is very easy to confuse them, originating an impedance mismatch. Furthermore, the dimensions of their inner conductors are slightly different, so mismatching the connectors may

FIGURE 4.22
Male type-N connector.

FIGURE 4.23
Female type N-connector.

FIGURE 4.24
Male and female subminiature version A (SMA) connectors.

also damage them physically. Subminiature version A (SMA) connectors, shown in Figure 4.24, are perhaps the most widely used microwave connector and are becoming more used in RFID systems. SMA connectors can transfer signals up to 18 GHz and have a 50 Ω impedance. The main problem using this type of connector is that they have not been designed to be connected and disconnected constantly, like for example a BNC connector. Even with the best of care, they cannot exceed a few hundred connections and disconnections.

4.4 Commercial Antennas for RFID Interrogators

4.4.1 Antennas for RFID Interrogators

Antennas for interrogators are available from the same manufacturers that make transponders and interrogators, as well as third-party vendors that build interrogator antennas for specific applications. Some manufacturers offer systems in which the antenna and the interrogator are integrated together, as is the case with handheld RFID systems.

Texas Instruments manufactures two types of antennas for the LF band under the generic name of Series 2000. These antennas are divided into *gate antennas* and *stick antennas*. Table 4.1 shows the specifications of these antennas.

As shown on Table 4.1, gate antennas mainly differ in their size. Designers of RFID systems choose the size of the antennas based on the required application, depending on the area that they have to cover. The stick antennas S01C and S02C are identical, with the only difference being the length of their coaxial cable: 1 meter for S01C and 3 meters for S02C. Choosing between gate antennas and stick antennas is also application dependent. In general, gate antennas offer a larger read zone and more read distance, while stick antennas offer a more focused zone. Therefore, if for example it is necessary to distinguish between different transponders, stick antennas will be more useful due to their more defined radiation field.

Some manufacturers have designed HF antennas that are shaped in a particular way for a specific application. For example, Unified Transponder Concepts, Inc., produces several antennas specific for laundry systems. One of them is shaped like a garment portal to be used with standard garment racks for clothes with an embedded RFID transponder. Another example from the same manufacturer is an antenna in the shape of a chute, used for the detection and control of deposited items tagged with the appropriate transponders. In addition, they also manufacture the more traditional panel and handheld antennas. The power levels accepted by these antennas ranges from 1 W to 4 W, and they all exhibit a resistance of 50 Ω at 13.56 MHz.

TABLE 4.1

Specifications for Some Texas Instruments Series 2000 Antennas Operating in the LF Band (134.2 kHz)

	Model	Inductance	Dimensions	Weight
Gate Antennas	G01E	27 µH	0.7 m x 0.2 m	745 grams
	G02E	27 µH	0.2 m x 0.2 m	425 grams
	G04E	26 µH	1.1 m x 0.52 m	2500 grams
Stick Antennas	S01C	27 µH	0.14 m x 2mm $(\varnothing)$	134 grams
	S02C	27 µH	0.14 m x 2 mm($\varnothing$)	185 grams
	P02A	116 µH	0.13 m x 21.3 mm ($\varnothing$)	105 grams

TABLE 4.2

Specifications for UHF Antennas Manufactured by NetHome

	Dimensions (mm)	Gain	Polarization	Connector
NT-900-SPP	90 x 90 x 25	0 dBi	Right / Left hand	SMA
NT-900-MPP	120 x 120 x 25	0 dBi	Circular	SMA
NT-900-CP	200 x 200 x 25	6 dBi	Right / Left had	SMA
NT-900-FP	360 x 260 x 25	6 dBi	Circular	TNC

TABLE 4.3

Specifications for MTI Wireless Edge Antennas Operating in the 400 MHz Range

Model	Frequency (MHz)	Dimensions (mm)	Gain	Polarization	Connector
MT182018	430 – 440	305 x 305 x 25	3.5 dBi	Circular LH	N-type Female
MT182016	430 – 437	370 x 370 x 30	9 dBi	Circular LH	N-type Female
MT181033	430 – 440	305 x 305 x 25	3 dBi	Linear	N-type Female
MT182011	405 – 450	370 x 370 x 40	8 dBi	Linear	N-type Female

TABLE 4.4

Specifications for MTI Wireless Edge Antennas Operating in the 865 MHz to 870 MHz Range

Model	Dimensions (mm)	Gain	Polarization	Connector
MT242032	190 x 190 x 30	7 dBi	Circular LH or RH	N-type Female
MT242040	190 x 190 x 30	7.5 dBi	Circular LH or RH	N-type Female
MT242033	540 x 470 x 220	8.5 dBi	Circular LH or RH	N-type Female
MT242034	685 x 626 x 220	8.5 dBi	Circular LH or RH	N-type Female
MT242027	260 x 260 x 30	8.5 dBi	Circular LH or RH	N-type Female
MT242014	305 x 305 x 25	8.5 dBi	Circular LH or RH	N-type Female
MT242017	371 x 371 x 40	10 dBi	Circular LH or RH	N-type Female
MT243017	650 x 320 x 40	11 dBi	Circular LH or RH	N-type Female

There is a larger number of independent manufacturers for antennas operating in the UHF range. For example, the Korean manufacturer NetHom offers flat, patch antennas that operate in the 902 MHz to 928 MHz range and have an input impedance of 50 Ω, as shown in Table 4.2.

Another manufacturer, MTI Wireless Edge Ltd., offers a vast array of standalone antennas for a wide number of frequencies and applications. Table 4.3 shows specifications for their line of antennas in 400 MHz UHF RFID range. All the antennas shown in Table 4.3 have a maximum input power of 6 W and an input impedance of 50 Ω.

The same manufacturer also offers antennas that operate in the 865 MHz to 870 MHz frequency range, have an input impedance of 50 Ω, and have a maximum allowable power of 6 W. Table 4.4 shows the rest of the specifications for these antennas.

TABLE 4.5

Specifications for MTI Wireless Edge Antennas Operating in the 2.4 GHz Range

Model	Dimensions (mm)	Gain	Polarization	Connector
MT343024	190 x 190 x 30	13 dBi	Circular LH or RH	N-type Female
MT344034	305 x 305 x 15	15.5 dBi	Circular LH or RH	N-type Female
MT345014	371 x 371 x 40	19 dBi	Circular LH or RH	N-type Female

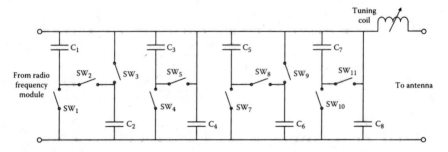

FIGURE 4.25
Schematic for the tuning board in the Series 2000 Reader from Texas Instruments.

It is also possible to purchase antennas similar to those shown in Table 4.4 dual polarized, that is, being both right-hand (RH) and left-hand (LH) circularly polarized. In this case, the antennas have two N-type female connectors, one for each polarization. Other models for antennas in the same frequency range of 865–870 MHz are available for linear polarization. In addition, this manufacturer also offers antennas for the 902–928 MHz and 950–956 MHz frequency ranges with dimensions and gains similar to those shown in Table 4.4 for either single or dual polarization.

Table 4.5 shows the specifications for a sample of antennas from MTI Wireless Edge Ltd. operating in the 2.4 GHz to 2.5 GHz frequency band and having an input impedance of 50 Ω.

Once again, antennas with similar dimensions and gains for the 2.4 GHz range are also available for single- and dual-linear polarizations.

4.4.2 Tuning Boards for Interrogator Antennas

Tuning boards are the circuits that are used to provide the high resonance needed to increase the strength of the field generated by the antennas in the system. Tuning boards can be found as stand-alone systems or as part of the interrogators. This subsection describes the Antenna Tuning Board that is part of the 2000 Series Reader System manufactured by Texas Instruments as an example of this type of circuits. As shown in Figure 4.25, the tuning board consists of a series of capacitors that are connected in parallel with the inductance of the antenna by means of several switches

TABLE 4.6

Resulting Capacitance for Different
Configurations of Switches in Tuning Board

Switch	Capacitance (nF)	Connection
SW1	33	C_2 ON
SW2	16.5	C_1, C_2 in series
SW3	33	C_1 ON
SW4	33	C_4 ON
SW5	16.5	C_3, C_4 in series
SW7	15	C_6 ON
SW8	7.5	C_5, C_6 in series
SW9	15	C_5 ON
SW10	10	C_8 ON
SW11	5	C_7, C_8 in series

(SW1–SW11) in the same board, thus allowing one to match a wide range of antenna inductances.

Table 4.6 shows the capacitive effect of each one of the switches in the tuning module. It can be seen that the lowest capacitance occurs when only SW11 is closed, resulting in a total capacitance of 5 nF. The highest capacitance that can be achieved with the board is equal to 139 nF, which occurs when SW1, SW3, SW4, SW7, SW9, and SW10 are closed.

The initial, coarse tuning adjustment is done with the capacitors, while the fine tuning due to the steps in the overall capacitance is done with the tuning coil located at the end of the tuning board. This process is repeated while monitoring the strength of the field generated by the antenna until it reaches a maximum.

5

Interrogators

CONTENTS

Interrogators have two basic functions: to generate and transmit the radiofrequency signal used to energize the transponders and to receive and decode the backscattered signal generated by the transponders. In addition, some interrogators may also transmit specific commands to the transponders. The interrogator also handles the bidirectional communication with a host computer used to process the information from the transponders and to issue commands to the interrogator. Figure 5.1 shows a basic block diagram of a generic interrogator with its basic constituent blocks.

If the transponder is able to accept commands from the interrogator, the radiofrequency signal generated by the interrogator has to be further modulated with the specific code; otherwise, the interrogator will only transmit the radiofrequency carrier. In either case, the radiofrequency signal is then amplified to specific power levels and passed through the tuning circuits before reaching the antenna. The antenna is also used to detect the backscatter signal generated by the transponder, which, after being filtered, is

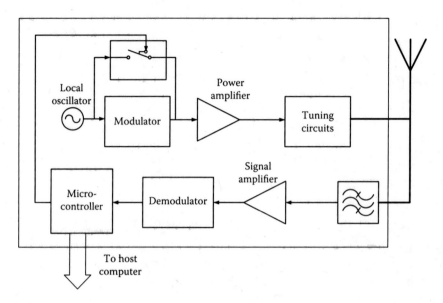

FIGURE 5.1
Basic block diagram of a generic interrogator.

then amplified and demodulated if necessary. The resulting signal is then sent to a microcontroller that will transmit it to the host computer using the appropriate communications protocol. The microcontroller also manages the timing for the interrogator to act as a transmitter or a receiver.

This chapter starts describing the transmitter and receiver sections in an interrogator. It continues by studying the most commonly used digital modulations in RFID systems. Afterwards, the chapter discusses issues related to extending the read range between transponder and interrogator as well as the requirements for interrogator synchronization when two or more interrogators coexist within the same general area. The last section in this chapter provides a description of the structure in different commercial interrogators.

5.1 Transmitter and Receiver

The basic structure of the transmitter section in the RFID interrogator contains an oscillator operating at the frequency of interest, a voltage or power amplifier, and the tuning circuits to match the impedance of the transmitter to the characteristics of the antenna. In addition to providing the radiofrequency signal to energize the transponders, some interrogators also transmit commands to the transponder. They can also be able to write new data into the transponder's memory. Transmitters in a RFID system must generate an accurate radiofrequency signal while minimizing the spurious radiation that they produce. *Spurious radiation,* common to all transmitters, is the amount of energy that falls outside the desired transmission band. In addition, they have to be as efficient as possible in order to reduce the amount of energy dissipated as heat. Excessive heat can severely reduce the life span of electronic components and must be reduced to acceptable levels.

For read-only transponders, the interrogator continuously transmits the radiofrequency signal and receives the response from the transponders. This mode of operation is known as *tag talks first.* When read/write transponders are being used, the interrogator must send commands for reading or writing them. For example, in the devices used by Microchip Technology, the interrogator uses amplitude modulation to transmit commands to the transponders. The communication starts by the interrogator sending specially timed gap pulses called *fast read request* (FRR) and *fast read bypass* (FRB). These gaps that are periods of time without electromagnetic field, are 175 µs wide, having a 100% modulation depth. The FRR or FRB pulses contain five of these gaps within 1.575 ms, as shown in Figure 5.2.

For RFID systems using magnetic coupling, one of the easiest approaches to drive the resonant circuit connected to the antenna is by using a D-drive circuit. Although either a full-bridge or a half-bridge will work correctly, most of the circuits use the half-bridge structure as it is less costly and easier to implement. Figure 5.3 shows a basic schematic for the half-bridge drive.

With a digital signal driving the circuit, the voltage at the node V_S is a square signal with voltages between V_{SS} and V_{DD}. Depending on the resonant quality factor (Q factor), the voltage across the antenna can reach a few hundred volts peak to peak. This is important to note at the time of selecting the electronic components used. An additional advantage of this structure is that although the driving digital signal has a very rich harmonic component,

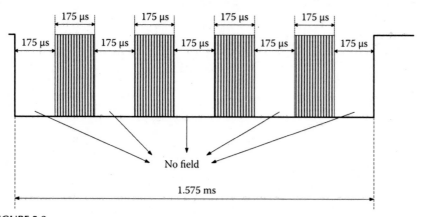

FIGURE 5.2
Fast read request (FRR) and fast read bypass (FRB) pulses.

the signal across the antenna consists mostly of the fundamental component due to the filtering effect and the high Q factor value at the antenna. There are several devices commercially available that integrate the inverter and the two metal oxide semiconductors (MOSs) or metal oxide semiconductor field effect transistors (MOSFETs) in a single package. In addition to being able to operate at the high voltages generated, the capacitors used must have very low tolerance values due to the requirements for an elevated Q factor.

The typical rise and fall times for the signals using the circuit shown in Figure 5.3 range between 120 µs and 200 µs. This limits the data transmission rate that can be achieved. A modification of the circuit to increase transmission rates by reducing rise and fall times needs to accelerate the turn-on or turn-off response of the driver. This can be achieved by using the modified driver circuit shown in Figure 5.4.

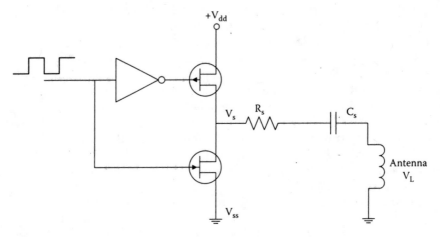

FIGURE 5.3
Half-bridge structure used to drive the resonant circuit.

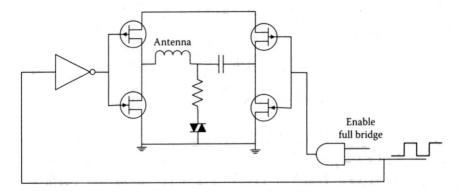

FIGURE 5.4
Modified bridge for increased transmission rates.

The idea behind this circuit is to decrease the start-up time by starting the driver in full-bridge mode until the voltage has reached the desired value, at which point it can be switched back to half-bridge mode. Because the driver is started by driving the two half-bridges 180° out of phase, it effectively operates like doubling the input source voltage. With this structure, the circuit can respond up to three times faster. It is also necessary to keep in mind that the voltage across the capacitor will also increase and therefore the capacitor must be rated for that higher voltage. In order to reduce the electromagnetic interference generated by the circuit, the Triac is typically fired at zero.

The basic structure of the receiver in the RFID interrogator also contains a detector, typically an envelope detector, a signal amplifier, filters, and the appropriate pulse-shaping circuits. These circuits convert the distorted received signal at the output of the filters into pulses. These pulses are, in turn, connected to the input of a microcontroller that will send the appropriate data to the middleware or host computers for further processing.

Selectivity and sensitivity are key parameters for any radiofrequency receiver. *Selectivity* is defined as the extent to which a receiver can differentiate between the desired signal and unwanted signals present in its bandwidth. *Sensitivity* is defined as the minimum input signal that produces the minimal required output to be interpreted as information. Other important characteristics for receivers are their dynamic range, cross-modulation, and intermodulation distortion. *Dynamic range* refers to the range of signal levels simultaneously applied to the input for which the receiver will operate correctly. *Cross-modulation* and *intermodulation distortion* originate as a result of the nonlinear effects in the receiver that cause it to create new, undesired components.

For RFID systems with magnetic coupling, the transponder communicates with the interrogator by changing the value of the magnetic field. The transponder is magnetically coupled to the front end of the receiver in the interrogator. In this structure, changes in the magnetic field become voltage changes at the antenna in the interrogator. Figure 5.5 shows a basic block diagram following this approach. The peak detector extracts the envelope of the signal being transmitted. The peak detector is followed by a filter to remove the DC component of the envelope. A further low-pass filter removes the radiofrequency component, leaving only the envelope of the modulated signal. The output of the comparator is a digital signal with the information from the transmitted signal.

The peak detector must operate with the relative high voltages that may appear at the input of the tuning circuit. Moreover, it should not become a load to the tuning circuit either. Other requirements for the peak detector are to reduce the ripple in the carrier as much as possible, maintain the modulation of the signal, and possess low setting times, among others. It is not possible to optimize all these parameters simultaneously, thus requiring the designer of the system to find an acceptable compromise depending on its overall requirements.

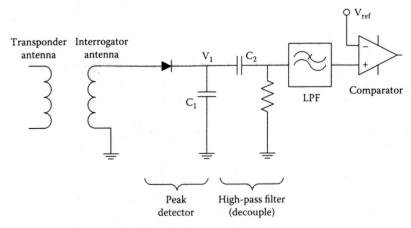

FIGURE 5.5
Basic structure for the detector in the RFID interrogator.

Example 5.1:

Discuss possible solutions for reducing ripple in the carrier.

Solution:

The ripple in the carrier can be reduced by increasing the value of the capacitor used in the peak detector. However, increasing the value of the capacitor will degrade the settling time for the detector, and will increase its loading effects. A possible solution would be to use an active peak detector. However, given the range of voltages expected at the input of the detector, the cost of the detector would significantly increase.

The voltage at node V_1 increases quickly when the input voltage increases due to the low impedance of the diode in forward mode. When the input voltage decreases, the voltage at node V_2 decreases extremely slowly due to the very high impedance presented by the diode in reverse conduction. This situation could be solved to some extent by placing a resistor in parallel with capacitor C_1 to help discharge it. However, this resistor has to be large enough for the peak detector to avoid loading the input circuits. A value of 10 MΩ is typically used.

The goal for the decoupling stage is to reject the high DC voltages present in the signal without excessive loading the front end. This stage should also have a fast dynamic response and stabilize quickly after detecting energy. The time constant of this stage is equal to the product of the resistance by capacitance C_2. The component values can be calculated by considering the delay time in establishing the communication after detecting the presence of radiofrequency energy as well as the minimum allowable voltage after the settling time. The cutoff frequency of the resulting high-pass filter has a strong effect in distorting and attenuating the digital signal that contains the information. With these conflicting requirements for the time constant, the simple RC circuit is not appropriate for removing the DC component of the received signal. Instead, it is necessary to use a circuit with either a nonlinear or a controllable time constant, as shown in Figure 5.6.

The diodes start conducting when the voltage across the resistor is above $(V_{REF} + V_{ON})$ or below $- (V_{REF} + V_{ON})$. This results in lowering the effective resistance seen by the circuit, and therefore it lowers its time constant. An additional benefit from using the diodes is that they protect the low-pass filter against positive or negative voltage spikes. The additional resistor R_2 is controlled by the switch. The switch is closed during transient periods and opened when data are being received by the circuit.

The low-pass filter stage is best implemented using an operational amplifier configured as an inverting amplifier, as shown in Figure 5.7. This structure not only allows controlling the gain in the pass band, but also minimizes its output impedance.

The cutoff frequency depends on the values of R_f and C_f. This frequency should be chosen at least two decades above the cutoff frequency from the DC blocker. It is also necessary to limit the gain of the filter to avoid the saturation of the operational amplifier. For some systems, the simple filter shown in Figure 5.7 may not provide enough attenuation at the desired frequencies. If this is the case, it can be substituted by a more complex filter. The three most common active filter topologies are Chebyshev, Butterworth, and Bessel filters. Chebyshev filters have the steepest transition between the pass and stop band, although they have ripple in the pass band. Butterworth filters have the flattest pass band, but their transition is not as steep as Chebyshev filters. Bessel filters have linear phase response with a smooth transition between

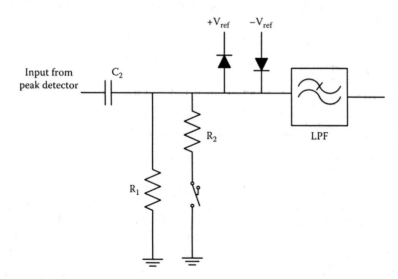

FIGURE 5.6
Circuit for implementing a nonlinear time constant.

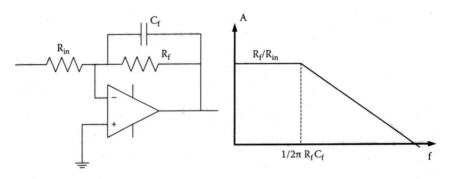

FIGURE 5.7
Low-pass filter stage.

pass band and stop band. When selecting one of these filter structures, it is necessary to consider not only their frequency response but also their transient response. For example, although Chebyshev filters have the best frequency response due to the fast transition between pass and stop bands, this results in an underdamped transient response with overshoot and ringing. Therefore, it has a very long settling time. Although Butterworth filters have a better transient response, they still exhibit some overshoot. The Bessel filter, although it has the worst frequency response, exhibits the best and fastest transient response and therefore is a very attractive candidate for this type of application.

The final stage is a comparator with some degree of hysteresis. The level of hysteresis is chosen to provide the required immunity against noise. The output of the comparator should be a digital signal that can be directly connected to the input of a microcontroller for further processing and transmission.

5.2 Modulator and Demodulator

5.2.1 Amplitude-Shift Keying (ASK) Modulation

Amplitude-shift keying (ASK) is the most basic form of modulation for digital signals. In ASK, data are represented as variations in the amplitude of the radiofrequency carrier. In binary ASK, there are only two symbols to be transmitted: the symbol *1* is normally represented by the full amplitude of the carrier, while the symbol *0* is represented by the minimal amplitude of the carrier. In this case, these two different symbols are represented by a single bit. In general, for *M*-ary ASK symbols, *n* bits are used to represent *M* symbols as

$$M = 2^n \tag{5.1}$$

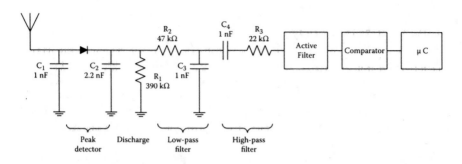

FIGURE 5.8
Detector in the receiver section of a commercial interrogator.

A M-ary ASK system requires distinguishing between M different amplitude levels for the radiofrequency carrier. While M-ary systems increase the transmission rate because each symbol encodes n different bits, the amplitude levels of the different symbols become closer to each other, making the system more susceptible to transmission errors. Therefore, there is a trade-off between the transmission rate and acceptable transmission errors.

The ASK signal is demodulated by detecting the envelope of the carrier, normally using a half-wave rectifier. This detects the peak amplitude of the signal generated by the transponder and feeds it into an RC circuit that charges and discharges accordingly. The time constant of the RC circuit must be chosen with these two considerations: (1) it must be small enough so the voltage across the capacitor diminishes fast enough in order to keep with the changes in the envelope of the signal, and (2) it must be high enough to avoid excessive ripple in the detected signal. After the envelope has been detected, the resulting signal is passed through a low-pass filter and a signal-shaping circuit and finally fed into a microcontroller. Figure 5.8 shows the schematic of the detector in the receiver section of a commercially available interrogator working in the LF range at 125 kHz.

The capacitor C_1 provides the resonance for the inductance of the antenna. An additional advantage of this structure is that because C_1 is grounded, the carrier finds a path to ground. The diode and C_2 form the half-wave rectifier for which the resistor R_1 provides a discharging path. The time constant of this circuit is approximately 850 µs. The signal is further low-pass and high-pass filtered before entering the comparator for processing.

5.2.2 Frequency-Shift Keying (FSK) Modulation

Frequency-shift keying (FSK) is a form of modulation in which the digital signal to be transmitted changes the frequency of the carrier within a predetermined set of available frequencies. The most basic case, in which only two instantaneous frequencies can be transmitted, is called *binary FSK* (2FSK). For more complex systems, and in order to increase the bit transmission rate, the number of possible instantaneous carrier frequencies is higher, normally a power of 2. This situation is called *M*-ary FSK, where *M* represents the total number of different possible instantaneous frequencies, as shown in equation (5.1).

Binary FSK transmits two different frequencies f_0 and f_1 to represent a logic 0 or a logic 1 following an established protocol. The transmission of 2FSK can be easily accomplished by having two oscillators operating at frequencies f_0 and f_1, followed by a switch that selects one or the other based on the digital input. Although this is a simple procedure, it results in discontinuities of the carrier frequency at the time of switching, thus resulting in a greater prominence of high-frequency components that implies the need for a much wider bandwidth transmission. 2FSK can also be generated using the VCO approach as is done for analog signals. Because the digital modulating signal will never have a zero rise or settling time, this digital modulation is in fact an analog modulation with a very short transition time. In this case, the carrier has a continuous and smooth transition, therefore guaranteeing phase continuity and requiring a narrower spectrum for transmission. This method is known as continuous-frequency shift keying (CPSK) and can also be readily applied for multilevel modulation systems.

FSK signals can be detected by *noncoherent detection* or *coherent detection*. Noncoherent detection is based on simultaneously injecting the 2FSK signal into two band-pass filters, each one tuned to the transmitted frequencies, f_0 or f_1, as shown in Figure 5.9. In this case, the output of each filter is connected to an envelope detector and their outputs are combined to reconstruct the baseband modulating signal.

An alternative method for noncoherent FSK detection is based on injecting the FSK signal to the input of a phase-locked loop (PLL). The output of the internal *voltage-controlled oscillator* (VCO) will switch between frequencies following the signal at the output of the PLL, thus locking on to the incoming FSK signal so that the output changes between the logic levels of 0 and 1, depending on the frequency of the incoming signal.

Figure 5.10 shows the basis for coherent detection of FSK signals. This approach requires the generation of two carriers with the same exact frequencies and phases as those used by the transmitter. In this case, only one

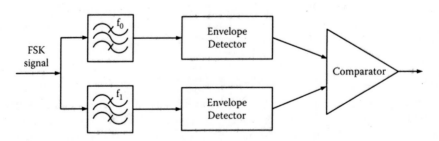

FIGURE 5.9
Noncoherent frequency-shift keying (FSK) demodulator.

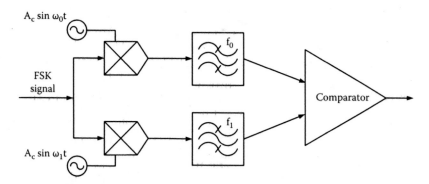

FIGURE 5.10
Coherent FSK demodulator.

of the filter outputs contains a signal at any given time. Therefore, after the addition of the two paths, the transmitting signal is recovered.

The coherent detection of FSK-modulated signals requires a more complex receiver than the noncoherent detection of the same signal, although it decreases the rate of errors for the transmission. Consequently, the designer of the system must consider the trade-offs between bit error rates and system complexity.

5.2.3 Phase-Shift Keying (PSK) Modulation

Phase-shift keying (PSK) modulation is one of the most effective methods to transmit digital signals. In PSK, the information is contained in the instantaneous phase of the modulated carrier. Usually, this phase is imposed and measured with respect to a known and selected phase of the carrier. For 2PSK, in which the modulating signal has two levels, the phase states are 0° and 180°. In *M*-ary modulation systems, the phase associated to each transmitted symbol is distributed equidistantly from each other around the 360° circle. When $M = 4$ ($n = 2$ bits), the resulting modulation is called *quadriphase phase-shift keying* (QPSK). QPSK, in which the phases are separated by 90°, is a common modulation used in communication systems.

Figure 5.11 shows a simplified version of one of the methods to generate a 2PSK signal. After the radiofrequency carrier is shifter by 180°, the resulting signal as well as the original signal are fed at the input of a 2-to-1 selector. This selector is driven by the digital signal. The output of the selector is either signal (a) or signal (b), depending on the value of the digital signal that controls the output of the selector.

PSK signals always require coherent demodulation, and therefore the PSK detector needs to know perfectly the phase of the unmodulated carrier. This requirement can be problematic as any phase error of the locally generated carrier reduces the effective voltage at the output of the detector in $\cos(\Phi)$, thus degrading the signal-to-noise ratio. A solution to this problem is by the

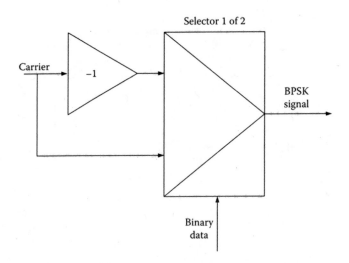

FIGURE 5.11
Binary phase-shift keying (2PSK) generation.

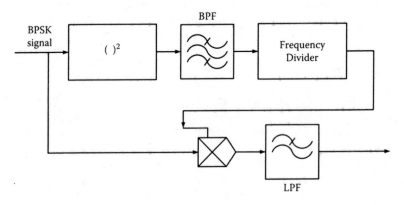

FIGURE 5.12
2PSK demodulation.

demodulator itself deriving a phase coherent reference to use in the demodulation process, as shown in Figure 5.12.

The output of the squaring circuit contains information at DC as well as at a double of the radiofrequency carrier. The band-pass filter eliminates the DC component. The frequency divider divides the frequency of the resulting signal by two and shifts its phase. The resulting signal is then multiplied by the original 2PSK signal that, after being filtered, will result in a DC signal with amplitude either positive or negative, thus shifted in 180°.

Figure 5.13 shows the generation process for QPSK signals. The digital-modulating signal $v_m(t)$ is connected to a 2-bit serial to parallel converter, so the resulting signals $v_1(t)$ and $v_2(t)$ change at half the rate of $v_m(t)$. This has the

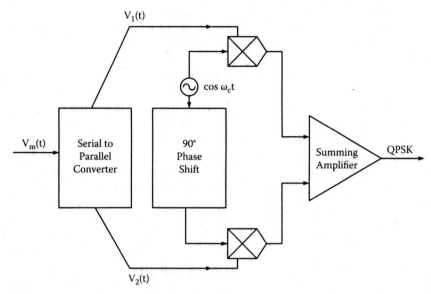

FIGURE 5.13
Quadriphase phase-shift keying (QPSK) generation.

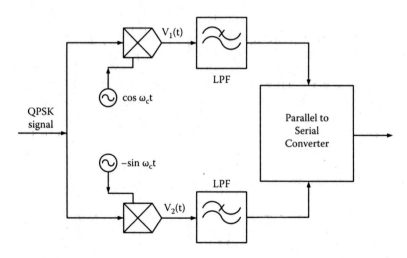

FIGURE 5.14
QPSK demodulation.

advantage of reducing the required transmission bandwidth as is the case of *M*-ary modulations. The output QPSK signal contains two components at the same frequency but shifted in 90°, and therefore there is no interference between them as they are in quadrature. That is, all the spectral components of the first term are 90° out of phase with those of the second term, making them independent from each other.

The detection of the QPSK signal, as shown in Figure 5.14, follows an inverse process. The intermediate output signals $V_1(t)$ and $V_2(t)$ contain energy at DC as well as a frequency double the radiofrequency signal. This last component is eliminated by the low-pass filter, resulting in two separate streams of digital data. There are combined into a single data stream by the parallel-to-digital converter, thus recovering the original digital-modulating signal. QPSK modulation allows for sending information at twice the speed of 2PSK using the same bandwidth without compromising the performance of the system from the point of view of error rates.

Figure 5.15 shows a basic diagram of a 2PSK demodulator used in a commercial RFID interrogator operating in the LF range at 125 kHz. Capacitor C_1 provides the capacitance for resonance with the inductance of the antenna. The two diodes are used for detecting the envelope of the signal; each diode is on during half of the cycle. The 1 MΩ resistor is used to discharge the voltage store in capacitor C_2 used for the envelope detection. The envelope signal is fed into a series of active filters and pulse-shaping circuits not shown in Figure 5.15. The output of the last pulse-shaping circuit is a square wave at a frequency that is half of the radiofrequency carrier signal. This signal exhibits a 180° phase shifts according to the digital data sent by the transponders. This signal is now used as the clock for a D-type flip-flop. The input to the flip-flop is a signal at the same frequency derived from the transmitting signal. As the phase of the detected signal V_{out} changes, the output of the flip-flop also changes, thus recovering the digital information sent by the transponder.

One of the major problems with the design shown previously is that the phase of the returned signal can be, in certain conditions, unpredictable. For example, if the transitions of the incoming signal and the reference signal occur at the same time, the output of the flip-flop becomes unpredictable. This situation can be avoided with additional circuitry that intentionally shifts the phase between the signals to 90°, making it ideal for PSK detection.

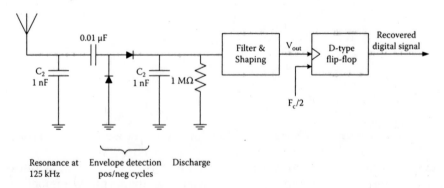

FIGURE 5.15
2PSK demodulator in a commercial interrogator.

5.3 Long-Range Considerations

When evaluating the read range of transponders, it is necessary to understand that long-range operation may not always be a desirable condition. For example, in an environment where several interrogators communicate with their own transponders, it is necessary to ensure that a specific interrogator reads data from only the transponders that it has been assigned. Data from a distant transponder would then be treated like an interfering signal.

If long range is truly needed, it can be achieved by one or more of the following design considerations:

Increasing the transmitter output power. This can be achieved by designing the power amplifiers in the transmitter section in the interrogator with higher gain. Alternatively, when it is not possible to modify the interior of the interrogator, it is then possible to use an external preamplifier. In any case, it is necessary to comply with the limits set by the appropriate regulatory agencies on limits of power for different frequency range. Also, some regulatory agencies may also limit the ability of a user to modify a system that has already been approved and certified by the regulatory agency. This comes from the fact that the modification of the gain in the amplifiers can have effects on the frequency stability, spectral purity, and other characteristics of the transmitted signal that are specified by the regulatory agencies.

Modifying the size of the antenna. As described in Chapter 4, the size of the antenna and its quality factor (Q factor) affect the characteristics of the signal being transmitted and therefore the range at which transponders receive enough energy to activate their internal circuits. A correctly sized antenna becomes critical for long-range applications.

Optimizing the sensitivity of the receiver. Merely increasing the power in the transmitter section may not have a significant effect in the read range of the system if the signals generated by the transponder cannot be detected by the receiver. Therefore, interrogators to be used for long-range applications must have front-end amplifiers with high enough sensitivity that will allow them to detect the signals from the transponders at the longest distance from the interrogator that has been specified. In addition to the optimal design of the receiver, the sensitivity of the interrogator can be increased by reducing the amount of noise and interference added to the desired signal. This can be achieved by appropriate shielding of the front-end components, by placing the interrogator away from electrically noisy sources and using pass-band filters.

5.4 Interrogator Synchronization

Synchronization is used to prevent interference between the interrogators in applications that have multiple interrogators operating in the same area. Synchronization is based on preventing interference by coordinating their transmission and reception windows. Synchronization is only required when the different interrogators are located physically close to each other. Therefore, when working with multiple interrogators, it is critical to determine if synchronization is required. However, the distance between interrogators alone is not sufficient to determine the need for synchronization. The electrical path between interrogators is affected by the presence of metallic structures such as buildings, the existence of conveyors, the layout of nearby power or data cables, and even the existence of reinforcing metal bars in concrete floors. Assuming only radiated signals, Texas Instruments has published some guidelines for the maximum distances that require synchronization for interrogators using some of their antennas as shown in Table 5.1. It is important to note that Table 5.1 assumes the two interrogators are using the same antenna type.

Table 5.1 shows that antennas with larger dimensions require more separation. Because this table assumes only radiated signals, in more realistic situations, this distance can be highly reduced due to the additional paths available to the different radiofrequency signals. This table can also be used as a starting point when working with interrogators using different antennas.

The next step, after determining that synchronization between interrogators is needed, is to decide on the method used to synchronize them. Although the following are the synchronization methods available to some of the Texas Instruments interrogators, the same methods are used by interrogators from other manufacturers.

5.4.1 No Synchronization at the Interrogator Level

This method can only be used with a single reader or in the case of synchronization by software. It represents the fastest method for receiving

TABLE 5.1

Separation between Interrogators After Which Synchronization Is No Longer Necessary Assuming Only Radiated Signals

Antenna Type	Antenna Dimensions	Minimum Separation Distance
S02-C (stick antenna)	$\varnothing$ = 21 mm; l =140 mm	18 meters
G02-E (gate antenna)	200 mm x 200 mm	32 meters
G01-E (gate antenna)	715 mm x 170 mm	55 meters
G04-E (gate antenna)	1018 mm x 518 mm	105 meters

information from the transponders. If all the interrogators are connected to the same network, the coordination can be carried out by the host computer issuing commands to read transponders simultaneously and buffering the results, or issuing different commands to different interrogators in turn. In any case, the coordination is done by the host computer rather than by the interrogators.

5.4.2 Wireless Synchronization

In wireless synchronization, when a specific interrogator detects increased energy at the frequency of the carrier, it assumes that this energy has been generated by another interrogator. This makes it stop itself before starting its own cycle. Obviously, this method only works when the background noise is low for the type of antenna being used. Larger antennas, which are more sensitive, present a clear disadvantage in this situation. When the environmental conditions change and there are additional reflections of signals in the immediate electrical environment, this approach is not valid either.

The receiver in each individual interrogator is adjusted in order to determine the level of background noise in the absence of signals from other interrogators. This determines the level after which the interrogator assumes that another interrogator is transmitting and therefore waits for a predetermined amount of time before starting its own cycle. This situation is illustrated in two situations depicted in Figure 5.16.

Figure 5.16(a) shows a time diagram involving wireless synchronization when the interrogator has not detected the presence of other signals. The interrogator starts by transmitting the radiofrequency signal for 50 ms that is used to energize the transponder. After the transponder is energized and recognizes the end of the radiofrequency energy burst, it starts transmitting its data for 20 ms. Once the transponder has finished sending its data, the interrogator waits for another 20 ms before starting the whole process again. This additional period is build as a precaution to avoid corrupting the data from other transponders to other interrogators. Because the interrogator does not detect any signals during this last 20 ms period, the whole cycle takes 90 ms.

In Figure 5.16(b), however, after the transponder has completed sending its information to the interrogator, the interrogator detects the presence of radiofrequency energy, and therefore responds by waiting 70 ms before starting its new cycle. This additional period gives enough time for the second interrogator to finish its own cycle. After the 140 ms have elapsed, the first interrogator starts its own cycle regardless of the presence of radiofrequency energy. This is done in order to prevent the interrogators always giving priority to other interrogators and therefore not being able to read their own transponders.

Wireless synchronization has the advantages that it does not need wires to interconnect the interrogators and all the interrogators are autonomous

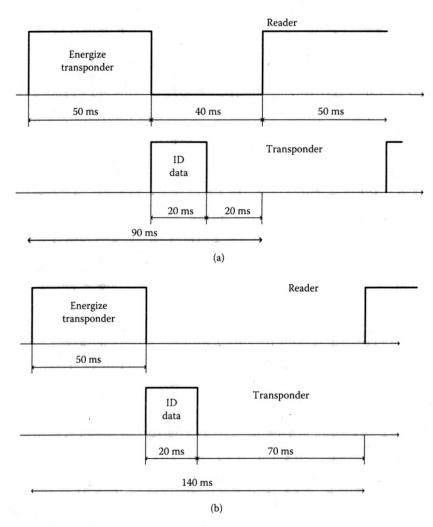

FIGURE 5.16
Wireless synchronization: (a) no other interrogators detected, and (b) delayed cycle due to other interrogators detected.

as there is not a master unit. It is very useful when handheld units have to coexist with fixed interrogators for these reasons. Wireless synchronization, however, is not adequate when large antennas are used in noisy environments because these large antennas are too sensitive to the background noise levels. Also, it cannot be used when the environmental conditions change, for example when vehicles enter or leave the read area as it changes the background energy. Finally, it cannot be used when other interrogators write data to the transponders.

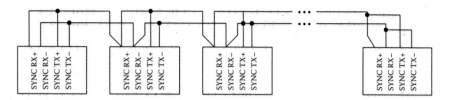

FIGURE 5.17
Structure of the control modules in interrogators using wired synchronization.

5.4.3 Wired Synchronization

The approach in wired synchronization is the same as for wireless synchronization with the only difference being that it uses wires to detect the presence of other interrogators instead of detecting radiofrequency energy. Therefore, the cycle for a single interrogator lasts 90 ms or 140 ms as described for wireless synchronization. This presents the advantage of being able to operate regardless of the environmental noise level, using a single twisted pair cable. This approach does not need a master unit. The main problem with wired synchronization is that if the power for a single interrogator fails, the bus fails and therefore the whole system fails. It is also not suitable for read/write transponders. Figure 5.17 shows the standard connection for the control modules of the interrogators using wired synchronization using the following nomenclature:

SYNC RX+: Noninverted data input

SYNC RX–: Inverted data input

SYNC TX+: Noninverted data output

SYNC TX–: Inverted data output

GND: Signal ground

5.4.4 Combined Wired–Wireless Synchronization

Using this approach, interrogators synchronized using a hard wire connection are also detecting the amount of radiofrequency energy to decide if another interrogator is already operational. This method resolves the individual disadvantages of each single approach.

5.4.5 Master–Slave Synchronization

In this approach, which is probably the most commonly used method of synchronization, one of the interrogators is configured as the master while the rest of the interrogators are configured as its slaves. Master–slave synchronization can be carried out using three variants:

Master–slave without acknowledgment. This method is the fastest approach for reading transponders as it assumes that all the interrogators are part of the same synchronization bus. This requires the interrogators to have a synchronization bus that connects all the readers, using for example a single twisted pair as shown in Figure 5.18.

Once the master has finished a complete read cycle, it transmits a synchronization pulse in the bus. Any slave that has finished its own cycle is waiting for this pulse in order to start a new cycle, as shown in Figure 5.19.

This approach presents the advantage of needing only a single twisted pair. It also has the fastest read rate and can be used when transponders are rewritten by interrogators. It is necessary, however, to keep in mind that if the master fails, the rest of the units stop working. Also, if a slave reads a transponder and the master does not, the slave may miss the next pulse as it is busy processing the information from its transponder.

Master–slave with acknowledgment. In this approach, the master has to wait until all the slaves have completed their cycles before starting a new master cycle. This requires using four wires: two for the pulse from the master and two for the information from the slaves to the master, as shown in Figure 5.20.

All the interrogators are running free while buffering the data via a point of data connection. A slave will not respond to a command from the host computer unless it has received the pulse from the master. This method presents the advantage that all the interrogators wait for the slowest to be complete and all the units are able to write to the transponders. This happens at the price of using a twin twisted pair instead of a single.

Master–slave triggered synchronization. This is similar to the previous approach, although in this case, the pulse is generated by an external device instead of an interrogator in the chain. All the interrogators

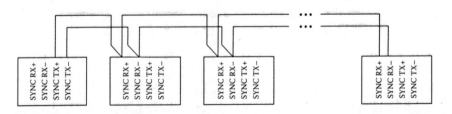

FIGURE 5.18
Structure of the control modules in interrogators using master–slave synchronization without acknowledgment.

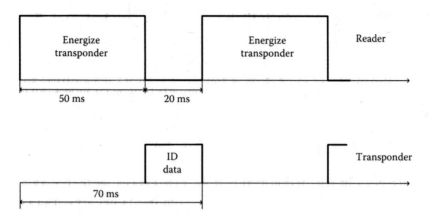

FIGURE 5.19
Master–slave synchronization without acknowledgment.

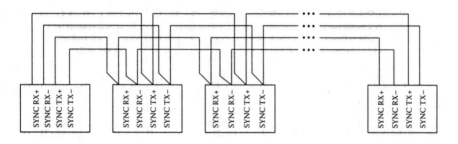

FIGURE 5.20
Structure of the control modules in interrogators using master–slave synchronization with acknowledgment.

are configured as slaves with the external master issuing the synchronization pulses at suitable intervals. This is required for interrogators that must read and write to multipage transponders in which different interrogators require different window times. In this case, the master has total control over the coordination of the devices in the chain. However, the master must be a more complex system and requires additional intelligence.

5.4.6 Carrier Phase Synchronization

This method is used when several antennas are located close together with the possibility that the field generated by each one of them may interact with the others. Because of the different transmission phases at each interrogator, the interaction may result in destructive interference. To avoid this effect, all the interrogators must be operated from the same carrier signal.

5.5 Structure of Commercially Available Interrogators

Interrogators for RFID systems take different approaches depending on the specific needs of their intended application: from large and complex systems to single-chip interrogators. The following pages describe these different approaches through examples from commercial interrogators. It is necessary to view them just as specific examples as the number of available systems is higher, each one of them using a different approach. However, these examples provide valuable information about the structure of interrogators.

5.5.1 Interrogators for LF RFID Systems

The Series Reader 2000 from Texas Instruments is an example of a complex interrogator for an RFID system operating in the low-frequency range. Its main component is the S251B reader. This interrogator has a communications interface as RS232, RS422, or RS485; eight general-purpose input/output lines; two open collector outputs; one synchronization bus; one carrier phase synchronization bus; as well as a power connector and an antenna connector. Table 5.2 shows the electric specifications of the eight general-purpose input/output (I/O) lines in this unit.

This interrogator supports two different communication protocols: the ASCII protocol or Texas Instruments' own TIRIS protocol. The ASCII protocol that is used to send commands to the interrogator can only be used with the RS232 or RS422 interface. The TIRIS bus protocol is a binary protocol between a host computer and one or more interrogators.

The interrogator has two main modules: the *control module* and the *radiofrequency module*. The control module contains all the electronics needed for the communication between the interrogator and the host computer in order to control the radiofrequency module. It also has circuits used to optimize the performance of the antenna by correctly tuning the module. The radiofrequency module handles all the analog functions needed to generate the

TABLE 5.2

Electrical Specifications of S251B Reader (Texas Instruments)

Parameter		Minimum	Maximum
General Purpose I/O Voltage			
	Low Level	---	0.9 V
	High Level	3.15 V	5.25 V
General Purpose I/O Current			
	Low Level	----	25 mA
	High Level	----	16 mA
General Purpose I/O 1 to 4 Total output current		----	10 mA
General Purpose I/O 5 to 8 Total output current		----	10 mA

signal that energizes the transponders, send commands to the transponders, and detect their responses. These two modules will be studied in deeper detail in Sections 5.5.1.1 and 5.5.1.2.

The synchronization bus operates in differential mode with a maximum data rate of 10 Mbits/s and allows for a maximum cable length of 1200 m. It can handle up to 32 different drives or receivers on the line. The outputs of the bus are protected against short circuits with a current limit of 150 mA to ground and 250 mA to the power supply. In addition, this unit contains three banks of dual in-line package (DIP) switches that allow for configuration changes. One of the banks is for setting up the control module, another bank for the settings of the communications protocol, and the third one for synchronization settings.

5.5.1.1 Radiofrequency Modules

The functions of the radiofrequency modules in the RFID system are to transmit the radiofrequency signal to the transponder via the antenna, to receive and demodulate the response of the transponder, and to write information to the transponder. Figure 5.21 shows a simplified block diagram of a generic radiofrequency unit in an interrogator.

For the modules that are part of the Series 2000 Reader manufactured by Texas Instruments, it is possible to distinguish the following submodules:

> *Power supply.* This submodule provides the necessary voltage to power the rest of the electronics in the radiofrequency module. The majority

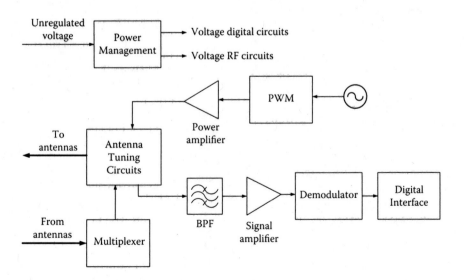

FIGURE 5.21
Simplified block diagram for the radiofrequency unit.

of modules contain two voltage regulators, one to power the logic circuits and the other to power the rest of the module.

The manufacturer stresses not using a switching power supply, as at their typical frequency of operation their harmonics will create strong interfering signals.

Transmitter. The radiofrequency signal is generated by a crystal oscillator operating at the RFID frequency of 134.2 kHz. This signal is then fed into the pulse width modulator for which the user can select the width ratio between 3% and 50%. It is necessary to consider that decreasing this ratio also decreases the energy of the transmitted signal, thus providing a method to control the power transmitted by the interrogator.

Antenna circuit. The purpose of this submodule, shown in Figure 5.22, is to match the impedance of the transmitter to the impedance of the antenna in order to maximize the radiated power as well as its Q factor.

The circuit consists of the inductance of the antenna in series with an adjustable inductance for fine tuning as well as the capacitive part formed by the capacitors C_1, C_2, C_3, C_4, and C_{couple}. The switch is used to connect or disconnect C_4 for coarse tuning. It is necessary to connect several capacitors in series because of the high voltages that can be developed across the antenna terminals during resonance.

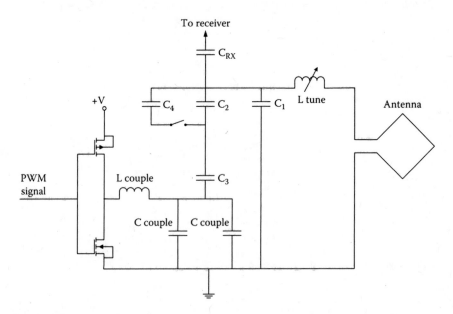

FIGURE 5.22
Antenna-tuning circuit.

The capacitive resonator is driven by the *pulse-width modulation* (PWM) signal from the transmitter by two MOSFET power transistors through the inductor L_{couple}.

Receiver. The receiver in this module demodulates the FSK signals generated by the transponders, with a low symbol specified by a frequency of 134.2 kHz and a high symbol by a frequency of 123.2 kHz. The module can operate with a single antenna used for transmission and reception or with specific transmission and reception antennas. The path for the received signal depending on the type of antenna used is selected by the multiplexer at the front end.

Table 5.3 describes the different pins and its connections for the standard radiofrequency module in the Series 2000 Reader from Texas Instruments. Modules from other manufacturers have similar connections.

The most important electrical characteristics for these pins and connectors are shown in Table 5.4.

As shown in Table 5.4, the radiofrequency module incorporates a signal strength detector whose output is the RXSS- pin. This allows adjusting the threshold level for the detection of radiofrequency energy required by wireless synchronization. The adjustment is done by means of a potentiometer on the board. Different types of antennas require different threshold levels even if they operate in the same environment.

TABLE 5.3

Pins and Connectors in Standard Radiofrequency Module

Signal Name	Connection	Description
GND	IN	Ground logic circuits
TXCT-	IN	Transmitter control input for activation
VSL	IN	Supply voltage
RXDT	OUT	Receiver data signal output
RXSA	IN / OUT	Adjustment for receiver threshold level
RXCK	OUT	Clock output receiver
GNDP	IN	Ground power stage
RXA0	IN	Select signal for multiplexer
GNDP	IN	Ground power stage
RDTP	OUT	Test pin for receiver
VSP	IN	Supply voltage power stage
RSTP	OUT	Test pin for receiver
VSP	IN	Supply voltage power stage
RXSS-	OUT	Signal strength output in receiver
VD	IN / OUT	Internally regulated voltage
RCTP	OUT	Test pin for receiver

TABLE 5.4

Summary of Electrical Characteristics Radiofrequency Module

Parameter	Min	Typ	Max	Unit
Internal regulated voltage	4.75	5.00	5.25	V
Supply current receive mode		9.0		mA
Supply current transmit mode		11.0		mA
Low level input voltage at TXCT- and RXA0	0		0.8	V
High level input voltage at TXCT- and RXA0	2.4		5.0	V
Low level output voltage at RXDT and RXCK	0		0.8	V
High level output voltage at RXDT and RXCK	4.0			V
Low level output voltage at RXSS-			0.8	V
Cable length for connection to control module		0.5	2.0	m
Number of slave modules that can be driven		1	5	
Inductance of antenna tuning coil	1.3	3.0	4.7	µH
Total antenna resonator capacitor	46	47	48	nF

This radiofrequency module, as well as most of the modules available in the market, can be used with the transmit–receive antenna or with special receive-only antennas. The type of antenna to be used is selected by a multiplexer in the receiver section.

5.5.1.2 Control Module

The control modules provide the functionality needed for the radiofrequency module to communicate with the rest of the devices. They control the Transmit and Receive commands from the radiofrequency modules according to the commands sent by the host computer. They also process the information from the transponders received by the radiofrequency modules, decode them, check their validity, and convert the signals to the protocol used for communication with the host computer. The control modules used in conjunction with the radiofrequency modules described in the previous section are available using the RS232C, RS422, or RS422/485 interfaces depending on the needs of the user. These control modules also handle the different synchronization methods described in Section 5.4 that need to be followed when more than a single interrogator is being used.

During the *charge time*, the control module activates the transmitter in the radiofrequency module. As long as this is active, the antenna resonates at the oscillator frequency and generates a radiofrequency field around its vicinity. This field charges the capacitor incorporated in the transponders energizing them. The typical length of the charge time is 50 ms. During this time, the receiver is disabled and cannot read any data. The charge time is followed by the *read time*, as shown in Figure 5.23. This time typically lasts 20 ms, during which the transponder sends the information back to the interrogator.

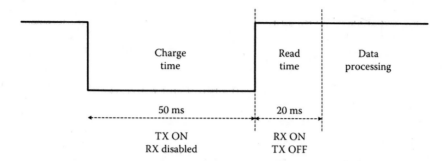

FIGURE 5.23
Charge and read times.

All the data formats consist of 128 bits. After being received by the radio-frequency module, they are checked by the control module. The different types of transponders supported by these interrogators are differentiated from each other by using different Start, Stop, and End bytes. The data structure for these transponders is shown in Figure 5.24. It is important to note that the first bit being transmitted is the least significant bit (LSB) and the last bit is the most significant bit (MSB).

The data structure for read-only transponders is shown in Figure 5.24(a). The data format starts with 16 pre-bits that are all set at zero, followed by the start byte ($7E). The next 64 bits are the identification data bits programmed into the transponder at the time of manufacturing; they cannot be changed by the user. The next 16 bits comprise the data protection bits (BCC). The following byte is the stop byte, which has the same structure as the start byte ($7E). Afterward, the transponder sends 16 end bits, all of them set to zero. During the last bit, the transponder discharges its capacitor, therefore deenergizing its internal circuits.

Figure 5.24(b) shows the data structure for read/write transponders. It also starts with the 16 pre-bits set at zero followed by the start byte that in this case is set at $FE. This tells the interrogator that the data come from a read/write transponder. The next 80 bits are the data bits that can be programmed by the user. The manufacturer recommends splitting these bits into 64 bits of data identification transmitted first and 16 bits for data protection bits (block check character, or BCC). The stop byte that is transmitted next has the same value as the start byte ($FE). The final 16 bits reflect the 16 LSBs of the read data that must be checked for validity by the control module. Once again, the last of these 16 bits makes the transponder discharge its capacitor and deenergize its internal circuits.

The data structure for the multipage read/write transponder is shown in Figure 5.24(c). The data start again with the 16 pre-bits set at zero, followed by the start byte with a value of $7E. This start byte is the same as for read-only transponders, meaning that the distinction between these two structures will be possible after the reception of the read address and read address frame

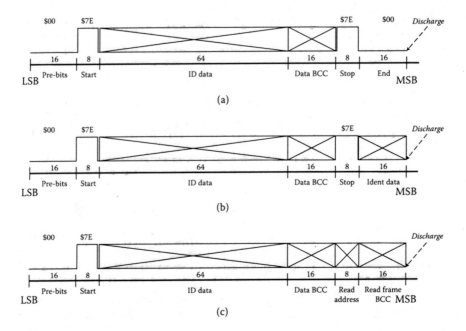

FIGURE 5.24

Data structure for several transponders: (a) read-only transponders, (b) read/write transponders, and (c) multipage read/write transponders.

BCC. The start byte is followed by 80 data bits that can be programmed by the user. The manufacturer also recommends splitting them into the first 64 data identification bits and 16 bits for protection data (BCC). The next 8 bits make up the read address byte; this contains a status field (2 bits), which is transmitted first, and a page field (6 bits). The 2 bits that make up the status field inform the control module about the function that the transponder has executed and the type of page that has been received. The final 16 bits make up the read frame BCC. One bit after having sent the first 128 bits, the transponder discharges the capacitor and deenergizes its internal circuits.

The function of the 16 pre-bits is to give enough time to the radiofrequency module to recover from the overload that it may have experienced during the 50 ms in which the transmitter has generated the radiofrequency signal. This is especially important because the transponders may start transmitting data immediately after having been energized. This period is known as the *waiting time*, lasting less than 1 ms.

After this time has elapsed, it is necessary to detect the first low-to-high transition for the start byte. After successful detection, the radiofrequency module synchronizes its clock and its data. The control module must also ensure that it receives all the high bits in the start byte. Otherwise, if a low bit is received, the process tops and the control module goes back to waiting for the detection of the low-to-high transitions. If the start byte has been successfully detected, the control module receives the next 80 bits and starts the generation of the cyclic redundancy check (CRC) to ensure the validity of the data. The last step in the decoding process is to determine whether the transponder that generated the data stream was a read-only transponder or a multipage read/write transponder as they both use the same start byte. This is done by checking the content of the CRC generator at the end of the process. If this value is equal to $0000, this indicates that a multipage read/write transponder originated the data. Otherwise, it indicates that it came from a read-only transponder. Figure 5.25 shows an overview of this sequence for the control module.

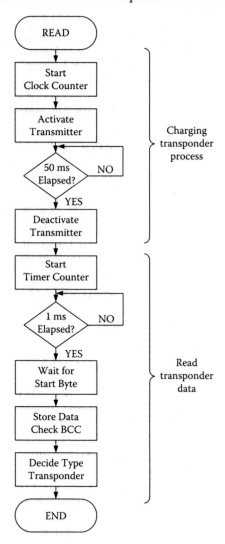

FIGURE 5.25
Control module sequence.

The control module is also used to control the flow for writing to the transponder. Writing to the transponder is used to transfer commands, addresses, and data to the transponder. Before writing to the transponder, this must be energized. Therefore, writing to the transponder occurs after the transmitter has sent the 50 ms radiofrequency pulse to energize the circuits in the transponder. The actual process of writing is done by switching the transmitter off and on. To write a high bit, the transmitter is inactive during 1 ms followed by being active during 1 ms. To write a low bit, the transmitter is inactive during 0.3 ms followed by being active during 1.7 ms. The total time for writing a single bit is 2.0 ms.

To program the write data into the transponder's memory, the transmitter must be active for at least 15 ms immediately after the previously described write function has been executed. This is known as the *programming time*. Having the transmitter on for this time enables it to energize the circuits in the transponder. The write data format is shown in Figure 5.26.

All the data bits are transmitted to the transponder with the LSB first. The 8-bit write keyword ($BB) and the 8-bit write password ($EB) are sent to the transponder to initiate the transmission of the 80-bit data. The manufacturer recommends splitting the 80-bit data into 64-bit identification data transmitted first followed by 16 protection data bits (BCC). The write data format is terminated by a 16-bit write frame that must be set to $0300. After write frame, the transmitter will be active for 15 ms to allow the transponder to program the received data into its memory. The actual programming will only take place if the write keyword and write password bytes are received, the write data have the correct number of bits, and the radiofrequency field generated by the transmitter is high enough to produce the voltage required to program the transponder's memory. This voltage exceeds the voltage required to read the memory. Once after the programming time has elapsed, the transponder responds by sending the data that it has stored into its memory. This data follow the 128-bit format described previously in this section. If the read data received are structurally valid, they must be compared against the write data that were sent to the transponder to ensure that the correct data were written into the transponder. If the structure or the contents of the read data are not valid, the process must be repeated.

The last main function for the control module is to handle the synchronization tasks between different interrogators. For this to occur, a synchronization

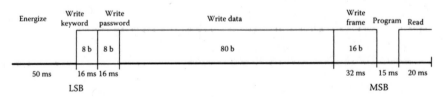

FIGURE 5.26
Format for writing data.

FIGURE 5.27
Synchronization sequence.

time must exist before the transmitter in the interrogator generates the radio-frequency signal to energize the transponders, as shown in Figure 5.27.

The synchronization allows the different interrogators in the overall system to detect each other and delay the activation of their own transmitters until other interrogators have finished their read functions. For this reason, the synchronization times vary depending on the number of interrogators in the application as well as the relative order of one interrogator in reference to the others in its vicinity.

5.5.2 Interrogators for HF RFID Systems

The Series Reader 6000 from Texas Instruments is an example of an RFID interrogator that operates in the HF range. The structure of the system is similar to the 2000 Series Reader described in the previous section. The Midrange Reader Module in this system provides the communication between transponder and interrogator. Output power levels can be adjusted between 100 mW and 1000 mW by means of a potentiometer placed on the radiofrequency module of the interrogator. The communication from the interrogator to the transponder is by ASK modulation, normally at 20%. The communication from the transponder to the interrogator is done by FSK modulation.

The TRF7960/1 also from Texas Instruments comprises multistandard fully integrated circuits containing the analog front end and the data-framing system. Their internal receiver enables them to modulate ASK and PSK, and it also includes automatic gain control and selectable bandwidth. In addition, the receiver includes a framing system allowing one to perform CRC checks, parity checks, or both. The output data are then sent to a microcontroller. The output of the transmitter only needs to be connected to the same impedance-matching circuit used for reception. Figure 5.28 shows the basic structure of the RFID system using this integrated circuit.

The receiver has two inputs that must be connected to an external filter. The external filter also converts the PSK modulation from the transponder into an amplitude-modulated signal. The two receiver inputs are multiplexed to two receiver channels: a main receiver and an auxiliary receiver. The main receiver contains a radiofrequency detection stage, a gain, a filter with automatic gain control, a digitizing stage, and an indicator for the strength of the received signal. The auxiliary receiver has a similar structure, although it is mainly used to measure the strength of the received signal.

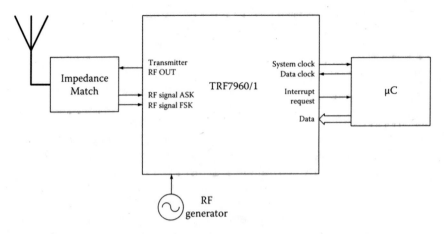

FIGURE 5.28
Structure of an interrogator using the TRF7960/1 circuits.

This is done by measuring the voltage of the demodulated signal and latching at its peak value.

The digital part of the receiver contains the bit decoders for the various protocols that supports the framing logic. The bit decoders convert the decoded signal into a bit stream. It has been specifically designed for maximum error tolerance, thus allowing the correct decoding of signals that have been partially corrupted due to noise or interference. The framing section packages the received bits into bytes, removing the control bytes sent by the transponder. This results in raw data that are sent to the microcontroller.

The transmitter contains the radiofrequency oscillator at 13.56 MHz, the processing circuits, and the radiofrequency power stage. The power levels available are 20 dBm (100 mW) or 23 dBm (200 mW) when the circuit is powered with 5V. When it is powered with 3 V, the available power levels are 15 dBm (33 mW) or 18 dBm (70 mW). The digital portion of the transmitter is very similar to that of the receiver as these circuits frame the command signals originated by the microcontroller.

5.5.3 Single-Chip Interrogators

In contrast to the complex systems described earlier, some interrogators have minimal components, normally an integrated circuit containing the transmitter and receiver and circuitry to communicate with a microcontroller or a microprocessor.

The core of the Series 6000 Reader from Texas Instruments is the RI-R6C-001 Integrated Circuit manufactured by the same company. This chip, whose block diagram is depicted in Figure 5.29, is a transceiver that provides the transmit and receive functions necessary for communication with transceivers operating at that frequency range.

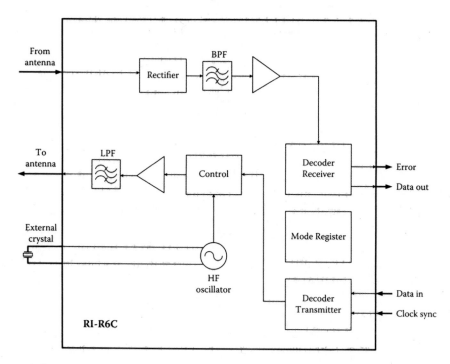

FIGURE 5.29
Simplified structure of the RI-R6C integrated circuit.

This transceiver integrated circuit communicates with its remote controller, normally a microprocessor, by a simple three-wire serial link. The communication between the transceiver and the transponders is generated by the microprocessor that sends the commands directly to the transceiver. The transceiver, in turn, applies the necessary carrier modulation and generates the radiofrequency signal.

The U2270B integrated circuit manufactured by Atmel® is another example of a single chip that serves as an interrogator. As shown in Figure 5.30, the U2270B incorporates an on-board power supply, an oscillator, tuning circuits, and other signal-processing circuits to produce an output that can be read by a microcontroller.

This specific chip has been designed to operate in the low-frequency range. It can be configured to work with different ranges of voltages for power supplies (from 5 V to 16 V). The low-pass filter in the receiver is a fourth-order Butterworth designed to remove the remaining energy of the carrier signal and other high frequencies after the demodulation process. The cutoff frequency can be selected by the user depending on the type of communication between the chip and the transponder that is being used. The possible choices are a Manchester code or biphase modulation typically at 5 kbaud.

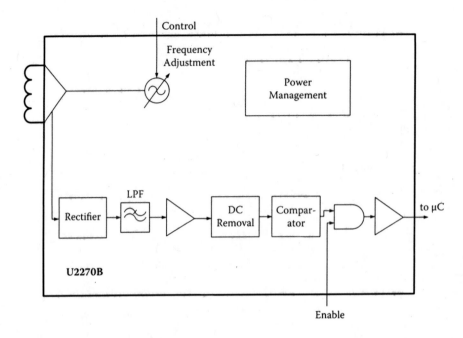

FIGURE 5.30
Simplified structure of the U2270B integrated circuit.

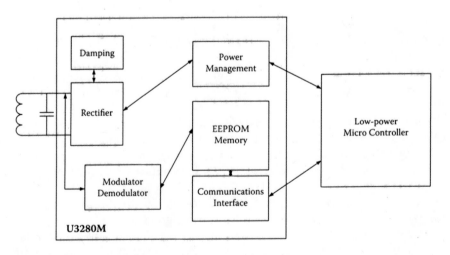

FIGURE 5.31
Simplified structure of the U3280M integrated circuit.

One of the main applications of this chip is as an electronic immobilizer for the automotive industry in order to reduce auto theft. The interrogator and the associated microcontroller are located inside the vehicle, while the transponder with its unique identifier code is carried by the lawful owner or user in the vehicle. Unless the interrogator receives a valid code from a transponder, the microcontroller keeps the electronic ignition of the vehicle disabled. An additional benefit of this approach is that because the power to the transponder is supplied by the interrogator through the electromagnetic field it generates, the user of the vehicle does not need to worry about carrying fresh batteries like in traditional wireless fobs.

Atmel® also manufactures another chip, the U3280M shown in Figure 5.31, which, although marketed as a transponder interface for microcontrollers, functions as an interrogator. The chip can supply the energy required to power an external microcontroller and can also accept an external battery. The distribution of power is controlled by a power management circuit that either operates in automatic mode or is controlled by the microcontroller. The communication of this device with its microcontroller is carried out following a serial format. Modulation is carried out by changing the magnetic field generated by the antenna by varying its load. For example, a high logic level results in an increase of current through the antenna while the voltage becomes damped. The low logic level is generated by a decrease of current through the antenna that results in an increase in its voltage.

6

Interrogator Communication and Control

CONTENTS

Interrogators communicate with the host computers in order to transmit commands to the transponders and collect their information. This chapter explores some of the protocols used for linking interrogators with the host computers using as examples the different protocols employed by some interrogators manufactured by Texas Instruments. These examples show the basic communication processes between interrogators and their host computers as a stepping-stone to understand the control of interrogators from other manufacturers. The chapter starts by describing the TIRIS Bus Protocol, developed by Texas Instruments. It continues by studying how the ASCII protocol is used in some commercial interrogators. The next section explains the protocol used in some LF interrogators manufactured by Texas Instruments. The chapter ends by describing the communication protocols used in interrogators operating in the HF range.

6.1 TIRIS Bus Protocol: Texas Instruments

The TIRIS Bus Protocol is a proprietary protocol developed by Texas Instruments for communications between its interrogators or between interrogators and a host. This is a half-duplex byte-count protocol. It allows efficient communication between a single master and multiple slaves. The protocol allows communication with a single slave unit on a RS232 line and up to 31 slaves on a RS422 or RS485 line or using parallel links.

All the data transfers using the TIRIS protocol are initiated by the host; the slave units only send data upon request from the host unit, therefore eliminating the possibility of data collision on the bus. Furthermore, the protocol uses positive acknowledgment with the slave units indicating the reception of a message from the master. All the messages have the same format independently of the direction they travel. Table 6.1 shows the structure of an *N*-byte message using this protocol.

Command-message-codes describe the action that the master requires from the slave. Its most significant bit (MSB; Bit 7) is the *queued-response flag*, while the following seven bits contain the command code. When the queued-response flag is set to 1, it indicates to the destination unit that it should acknowledge and execute the command but its response should be queued until requested by the master. This is normally used with commands that have a delayed response in order to avoid suspension of the line activity. When the flag is set to zero, the response is immediate following the execution of the command.

TABLE 6.1

TIRIS Bus Protocol Format

Byte	Content
0	Start Mark[a]
1	Destination Address[b]
2	Source Address[c]
3	Message Code[d]
4	Data length
5	Data field (1)
6	Data field (2)
...	...
N + 4	Data field (N)
N + 5	MSB for CRC-field (1)
N + 6	CRC – field
N + 7	End Mark

[a] The *start mark* signifies the beginning of a message and is represented by $01. This corresponds to the ASCII character Start-of-Heading.

[b] The second byte, *destination address*, indicates the address of the unit to which the message is directed. The binary value of this byte corresponds to the ID of the destination address. The ID for the different units ranges from $00 to $FE, although the maximum number of units in a specific application is limited by the physical limitations of the link. Broadcast messages are indicated by using a destination address of $FF.

[c] The *source address* is the third byte in the message format. It indicates the ID of the unit sending the message.

[d] The fourth byte is the *message code* byte. This byte is used to define the content of the message. When the message is from the master to the slave, this byte is a *command-message-code*. When the message is from slave to master, this byte is a *response-message-code*.

The response-message-code describes the acknowledgment that the slave unit has received the command. The four MSBs in the code are flags, while the four remaining least significant bits (LSBs) are the response code. The different flags are as follows:

The MSB (Bit 7) is the *error-flag*. When it is set to 1, it indicates that the slave has detected an error in the transmission or in the contents of the command. The error code is indicated in the last four bits.

Bit 6 is the *busy-flag*. When it is set to 1, it indicates that the slave is temporarily unable to accept commands.

Bit 5 is the *data-available-flag*. When it is set to 1, it indicates to the master that there is at least one message in the output queue in the slave. In order to read this message, the master must send the appropriate command.

Bit 4 is the *broadcast-received-flag*. When it is set to 1, it indicates that a broadcast message, issued previously, has been received correctly. The response to the broadcast command is placed in the queue. The master can read the response by issuing the appropriate commands at which point the flag is reset to 0.

Bits 5-8 in the Response-Message-Code byte indicate the type of response. Their meaning varies depending on the state of the Error-Flag bit as shown in Table 6.2.

Bit 5 in the TIRIS protocol is the *data length* byte. This byte shows the number of bytes that follow as data field. If the command or response does not require any data, this value will be zero.

The following N bytes in the TIRIS protocol make up the *data field*, with N being the value specified in the previous byte (Data Length byte). The content of this field depends on the message being sent.

The two bytes after the Data-Field bytes are *CRC-field* bytes. These bytes contain the Cyclic Redundancy Check values to ensure the integrity of the message. The TIRIS Bus protocol supports two methods of generating the CRC field. The first method is the Reverse CRC-CCITT algorithm that was described in Section 3.3.2. The second method supported by this protocol is called Longitudinal Redundancy Check (LRC). This is based on XORing each character with the previous value and storing the result in the LSB byte fin the CRC-Field. The MSB byte of the CRC-Field is generated by storing the 1's complement of the LSB byte. Using this method, if the MSB and LSB are ANDed together, the result should be zero. Consequently, any value different from zero indicates an error in the received message. This method is not as secure as the Reverse CRC method but it is faster.

The final byte is the *end mark* byte that indicates the end of the message. This is represented by $04 that corresponds to the ASC II character *end of transmission*.

TABLE 6.2

Response Codes

Bit	Error-Flag = 0 Process Response		Error-Flag = 1 Error-Response	
0	Command Complete	A non-queued command received and executed correctly	Transmission Error	Slave received a message with its ID but found error on data
1	Accepted	A queued response command received correctly. Response will be placed on the queue	Command Invalid	Message received correctly but command code is not valid for this slave
2	Queue Empty	No message in the queue	Error – Data field length	Data field length wrong. Command not executed
3	Nothing to Resend	No information to resend available	Error - Parameter	Invalid parameter. Command not executed

Note: The following *N* bytes make up the *data field*, with *N* being the value specified in the previous byte (data length byte). The content of this field depends on the message being sent.

Using a two-wire, half-duplex circuit is not possible to execute flow control. Therefore, all the units in the bus must be able to receive and buffer the maximum length of the message at the operating speed for the line. When a unit is unable to handle correctly the incoming data, it cannot send an error message until the original message has ended. At that point, it can transmit the appropriate error message.

The data flow for the commands that do not require queuing from the slaves is always controlled by the master. The first steps for the master unit are to enable the transmitter, send the command message, and disable the transmitter. The slave unit, which is always monitoring the line, receives the incoming message and checks for the correct destination address. If the destination address is incorrect, it ignores the message and returns to monitoring the line. Because in this case the slave does not give any response to the master, the master will resend the command.

If the destination address is correct, the slave unit checks for errors in the message. If it detects errors, it sends an error response to the master, disables its transmitter, and returns to monitoring the line. Upon reception of the error response, the master unit processes the error code and resends the original command. If no errors are detected, the slave unit processes the command, sends the appropriate response message, disables the transmitter, and returns to monitoring the line. Upon reception of the response message, the master processes it and proceeds to transmitting the next command.

Broadcast messages transmitted by the master are not acknowledged in the normal way. After the master has sent the broadcast command, the slave checks for errors in the received message. If it detects errors, it ignores the message, sets the error flag, and returns to monitoring the line. If it does not detect errors, it sets the broadcast-received flag, processes the command, and places the response in its queue. The master accesses the special status bit in the response-code field in the slave's response to evaluate the correct reception of the previous message.

Example 6.1:

Describe the operations taken by a master unit that wants to initiate an operation on all slaves simultaneously.

Solution:

The master unit sends a command with a destination ID of $FF. The master unit then waits the expected operation time and proceeds to poll each individual slave. The master is looking for the *broadcast-received-flag* that verifies the correct reception of the command. The absence of this flag in a slave unit triggers the master to broadcast the command again.

Queued messages use a slightly different format from the format used by broadcast or immediate commands. The Nth data field byte contains a sequence number, given by the master, that is used to have a reference when the queued messages are requested from the slave units. The slave unit indicates that a command requiring a queued response has been received and is being processed by sending a response message to the master with the accepted flag set. This response is not transmitted directly by the slave unit, but it is placed on its queue for the master unit to access it when ready.

When interrogators are asked to queue a response, these responses are placed in a 30-position circular buffer. The buffer has two pointers: store-next-message and read-next-message. Because of the circular structure of the buffer, it is not necessary to delete responses as they are overwritten when the pointer is incremented.

6.2 ASCII Protocol for TIRIS Interrogators: Texas Instruments

Some of the interrogators from Texas Instruments accept commands using a simple ASCII protocol. This ASCII protocol can only be used with a RS232 or RS422 communications line. Once the interrogator reinitializes, the first character to be transmitted is $02 (Start of Text, or STX), followed by $0D (Carriage Return) and $0A (Line Feed) through the serial interface. Any characters that may have been transmitted before the STX character are treated as random characters generated by the start-up process and therefore discarded.

Using this protocol, the interrogator can operate on four different modes: execute, normal, line, and gate. In execute mode, the interrogator triggers a single command, while in the other modes the read function is continuously triggered. The interrogators are also able to distinguish between three types of transponders: read-only, read/write, and multipage. The serial communications protocol from the interrogator to the host computer is introduced as a single character that describes the type of interrogator that originated the message: R (read-only), W (read/write), or M (multipage).

After receiving data from a transponder, the interrogator performs a data check. The result of this data check is one of the following: *identification received correctly, identification invalid,* or *no identification detected.* After receiving a correct identification, the interrogator proceeds to send the information through the bus. This is done by placing the identification string in a buffer that is accessed, read, and cleared by the host computer. The TIRIS protocol can distinguish between two groups of transponder identification: animal code numbers and industrial (nonanimal) code numbers.

6.3 Series 2000 Micro-Reader System: Texas Instruments

The Series 2000 RFID interrogators operate in the LF range. Table 6.3 describes the structure of an *N*-byte message sent from the host computer to the interrogator in these systems.

The following set of bytes are the *data field* bytes. These bytes only exist if required by the command field.

TABLE 6.3

Series 2000 Host-to-Interrogator Message Format

Byte	Content
0	Start Mark[a]
1	Message length[b]
2	Command Field[c] (1)
3	Command Field (2) - not always present
4	Data field (1)
5	Data field (2)
...	...
N + 3	Data field (N)
N + 4	BCC

Note: If the second command field is not used, the length of the message is *N* + 3 bytes. In addition, messages are limited to a total of 41 bytes.

[a] The first byte is the *start mark* that indicates the start of a message.
[b] The second byte is the *message length* byte that indicates the length, in bytes, of the command and data fields that follow.
[c] The third and fourth bytes (when the second command field byte exists) are the *command field* bytes. These bytes define the mode of operation for the interrogator and determine the operation that the interrogator must carry out on the transponder. Depending on the type of command, the information in the data fields may or may not be transmitted to the transponder. The meaning of the different bits for the command field (1) is described in Table 6.4.

TABLE 6.4

Command Field (I) Bit Description

Bit	Description	Value	Operation
0 and 1	Operation Mode	00	Perform single command
		01	Read in continuous Normal Mode
		10	Read in continuous Line Mode
		11	Send interrogator software version
2	FBCC Calculation	0 or 1	1 = interrogator calculates FBCC
3	Power burst I	0 or 1	1 = duration determined in data field
4	Power burst pause	0 or 1	1 = duration determined in data field
5	Power burst II	0 or 1	1 = duration determined in data field
6	Data	0 or 1	1 = determined in data field
7	Command Expansion Field	0 or 1	1 = Command Field (2) is used

TABLE 6.5

Command Field (2) Bit Description

Bit	Description	Value	Operation
0	Special write timing	0 or 1	1 = Must be determined in data field.
1	Wireless synchronization	0 or 1	1 = Wireless synchronization being used.
2	Database consistency check (DBCC) calculation	0 or 1	1 = Interrogator calculates DBCC.
3 to 7	Reserved	0 or 1	Reserved.

The following set of bytes are the *data field* bytes. These bytes only exist if required by the command field.

The last byte in the message is the *block check character* (BCC) byte. This is a one-byte value that results from XORing each one of the bytes in the preceding message with the exception of the start mark byte.

As shown in Table 6.4, the existence of command field (2) is set by bit 7 of command field (1). Table 6.5 shows the meaning of the bits in command field (2) when it is present in the message.

Example 6.2:

Calculate the BCC for a message containing the following hex bytes: 04 23 38.

Solution:

BCC is calculated by XORing the bytes of the message. Start by first XORing 04 and 23:

04	0000 0100
23	0010 0011
XOR	0010 0111

The resulting value (0010 0111) is now XORed with 38:

Previous XOR	0010 0111
38	0011 1000
XOR	0001 1111

Therefore, the resulting BCC is $1F.

The following three examples illustrate the structure of messages between the host computer and the interrogator for a variety of different commands and applications.

Example 6.3:

The following is the structure of a message transmitted from the host computer to the interrogator in commanding the interrogator to read a read-only transponder:

Byte	Code	Content	Description
0	$01	Start Mark	Start Mark
1	$02	Length	Two bytes follow (excludes BCC)
2	$08	Command Field (1)	Perform single command – Power burst
3	$32	Data Field (1)	Power burst for 50 ms to energize transponder
4	$38	BCC	BCC over previous bytes excluding Start Mark

Example 6.4:

The following is the structure of a message transmitted from the host computer to the interrogator in commanding the interrogator to write the following data in a read/write transponder.

00 0A	FE	32	66	00	00	01
MSB						LSB

Byte	Code	Content	Description
0	$01	Start Mark	Start Mark
1	$11	Length	17 bytes follow (excludes BCC)
2	$E8	Command Field (1)	Perform single command, no FBCC calculations, send power burst to energize transponder for reading and writing. Data command on following field
3	$06	Command Field (2)	Use wireless synchronization
4	$32	Data Field (1)	Power burst to energize transponder for reading (50 ms)
5	$0F	Data Field (2)	Power burst to energize transponder for writing (15 ms)
6	$0C	Data Field (3)	12 data bytes follow
7	$BB	Data Field (4)	Write Keyword
8	$EB	Data Field (5)	Write Password
9	$01	Data Field (6)	LSB of data to write
10	$00	Data Field (7)	Data to write
11	$00	Data Field (8)	Data to write
12	$66	Data Field (9)	Data to write
13	$32	Data Field (10)	Data to write
14	$FE	Data Field (11)	Data to write

Byte	Code	Content	Description
15	$0A	Data Field (12)	Data to write
16	$00	Data Field (13)	MSB of data to write
17	$00	Data Field (14)	Write frame
18	$03	Data Field (15)	Write frame
19	$9C	BCC	BCC over previous bytes excluding Start Mark

Example 6.5:

The following is the structure of a message transmitted from the host computer to the interrogator in commanding the interrogator to read page 2 of a multipage transponder.

Byte	Code	Content	Description
0	$01	Start Mark	Start Mark
1	$04	Length	Four bytes follow (excludes BCC)
2	$48	Command Field (1)	Perform single command – Power burst with data
3	$32	Data Field (1)	Power burst for 50 ms to energize transponder
4	$01	Data Field (2)	One data field follows
5	$08	Data Field (3)	Write address specifying general read of page 2
6	$77	BCC	BCC over previous bytes excluding Start Mark

The structure of the messages sent from the interrogator to the host computer is described in Table 6.6.

The bytes following BCC are the *data* field bytes.

TABLE 6.6

Series 2000 Interrogator-to-Host Message Format

Byte	Content
0	Start Mark[a]
1	Message length[b]
2	Status[c]
3	Data field (1) LSB byte
4	Data field (2)
...	...
N + 2	Data field (N) MSB byte
N + 3	BCC

[a] The first byte is the *start mark* that indicates the start of a message.

[b] The second byte is the *message length* byte that indicates the length, in bytes, of the command and data fields that follow.

[c] The third byte is the *status byte*. This provides feedback about the preceding read or programming operation according to Table 6.7.

TABLE 6.7

Description of Bits in the Status Byte

Bit	Value	Operation
0 and 1	00	Read Only transponder
	01	Read/Write transponder
	10	Multipage transponder
	11	Other transponder
2	0 or 1	1 = Start byte was successfully detected
3	0 or 1	1 = DBCC was successfully performed
4	0 or 1	1 = DBCC was successfully performed
5	0 or 1	1 = Software version for interrogator follows
6–7	Reserved	

TABLE 6.8

Description of Data Field Bytes

Type of Response	Number of Bytes in Data Field	Description
Read Only	8	Identification data. LSB first
Read/Write	8	Identification data. LSB first
Multipage	9	Identification data. LSB first and Read Address
Other	14	Complete transponder protocol without pre-bits assuming a successful start byte was detected
No Read	0	No data fields, no start byte. Status is $03
Software version	1	Software version. For example $12 means software version 1.2

The last byte in the message is the *BCC* byte. This is a one-byte value that results from XORing each one of the bytes in the preceding message with the exception of the start mark byte.

The length and meaning of the data field bytes are shown in Table 6.8.

Example 6.6:

The following is the structure of a message transmitted from the interrogator to the host computer after a successful read of a read-only transponder.

Byte	Code	Content	Description
0	$01	Start Mark	Start Mark
1	$09	Length	Nine bytes follow (excludes BCC)
2	$0C	Status	Valid RO, Start byte detected, DBCC is correct
3	$6A	Data Field (1)	Identification data (LSB)
4	$58	Data Field (2)	Identification data

Byte	Code	Content	Description
5	$4C	Data Field (3)	Identification data
6	$00	Data Field (4)	Identification data
7	$00	Data Field (5)	Identification data
8	$00	Data Field (6)	Identification data
9	$00	Data Field (7)	Identification data
10	$00	Data Field (8)	Identification data (MSB)
11	$7B	BCC	BCC over previous bytes excluding Start Mark

Example 6.7:

The following is the structure of a message transmitted from the interrogator to the host computer after an unsuccessful read attempt.

Byte	Code	Content	Description
0	$01	Start Mark	Start Mark
1	$01	Length	One byte follows (excludes BCC)
2	$03	Status	Other, no Start byte detected, DBCC is not correct
3	$02	BCC	BCC over previous bytes excluding Start Mark

Example 6.8:

The following is the structure of a message transmitted from the interrogator to the host computer after successfully programming page 2 of a multipage transponder.

Byte	Code	Content	Description
0	$01	Start Mark	Start Mark
1	$0A	Length	Ten bytes follow (excludes BCC)
2	$1E	Status	Valid Multipaage, Start byte detected, DBCC is correct
3	$47	Data Field (1)	Identification data (LSB)
4	$C6	Data Field (2)	Identification data
5	$2D	Data Field (3)	Identification data
6	$00	Data Field (4)	Identification data
7	$00	Data Field (5)	Identification data
8	$00	Data Field (6)	Identification data
9	$00	Data Field (7)	Identification data
10	$00	Data Field (8)	Identification data (MSB)
11	$09	Data Field (9)	Read address specifying successful programming of page 2
12	$7B	BCC	BCC over previous bytes excluding Start Mark

6.4 High-Frequency Interrogators: Texas Instruments

6.4.1 TIRIS Protocol for Series 6000 Reader System Interrogators

The Series 6000 Reader System interrogators are RFID interrogators operating in the HF band, manufactured by Texas Instruments. These interrogators use a RS232 line to communicate with the host computer. The data packet from the host to the interrogator is known as the *request*, while the data from the interrogator to the host are known as the *response*. All communication sequences are initiated by the host that is the primary station. Using this protocol, the host waits for the response before continuing, thus assuring that the host does not have to handle large sequences of open requests. This is a binary, byte-count-oriented protocol in which the data length is passed as a parameter in the message.

Table 6.9 shows the structure of the messages used by these interrogators for the request message (host to interrogator).

The contents of byte 5 (command flags) are used to control the actions of the interrogator. The meanings of the different bits in this byte are shown in Table 6.10.

The BCC, used to ensure the integrity of the message, is a 16-bit code calculated on all the bytes of the packet including the start mark. The BCC contains two parts: its least significant byte is a longitudinal redundancy check (LCR) calculated by performing a cumulative XOR operation on all the bytes of the packet. The most significant byte in the BCC is the one's complement of the LCR.

The structure of the message sent from the interrogator to the host (response) is shown in Table 6.11. Most of these bytes are similar to the bytes in the request message.

The data field in response messages is limited to a maximum of 23 bytes. The contents of byte 5 (command flags) are used to report on the actions of the interrogator. The meanings of the different bits in this byte are shown in Table 6.12.

The different error codes that the interrogator transmits to the host computer when bit 4 of the command flag byte is set to 1 are described in Table 6.13.

The contents of byte 5 (command flags) are used to control the actions of the interrogator. The meanings of the different bits in this byte are the same as for the request message shown in Table 6.10.

6.4.2 Host Protocol for Tag-it™ Series 6000 Reader System Interrogators

This protocol defines the communications between the Tag-it™ Series 6000 Reader System Interrogators and the host computer. It handles the communication between the host and the interrogator, and carries out the requests

TABLE 6.9

Series 6000 Host-to-Interrogator Message Format

Byte	Content
0	Start of Frame[a]
1	Message length[b] - LSB
2	Message length - MSB
3	Node address[c] - LSB
4	Node address - MSB
5	Command flags[d]
6	Command[e]
7	Data[f]
8	Data
...	Data
N+6	Data
N+7	BCC[g]
N+8	BCC

[a] The first byte is the *start mark* that indicates the start of a message. This byte is equal to $01.

[b] The second and third bytes are the *message length* bytes that indicate the length of the whole packet, including the start mark.

[c] Bytes 3 and 4 constitute the *node address*. Its value is $0000.

[d] Byte 5 is the *command flags* byte. It controls the actions taken by the interrogator. The value of this byte is dependent on the command being transmitted.

[e] Byte 6 is the *command* byte. It specifies the actions that must be taken by the interrogator. The value of this byte is dependent on the command being transmitted.

[f] The following N bytes are the *data* bytes. These contain the parameters and the data for the command.

[g] The last two bytes are the *BCC* bytes used to ensure the integrity of the preceding message. BCC is performed on the previous bytes, including the start mark.

TABLE 6.10

Bit Functionality in Command Flag Byte (Request)

Bit	Description
0	Reserved for future use
1	Reserved for future use
2	Reserved for future use
3	Reserved for future use
4	1 = Command to be performed only on transponders with an address that matches the data section of the message
5	Reserved for future use
6	Reserved for future use
7	Reserved for future use

TABLE 6.11

Series 6000 Interrogator-to-Host Message Format

Byte	Content
0	Start of Frame
1	Message length - LSB
2	Message length - MSB
3	Node address - LSB
4	Node address - MSB
5	Command flags
6	Command
7	Data
8	Data
...	Data
N+6	Data
N+7	BCC
N+8	BCC

TABLE 6.12

Bit Functionality in Command Flag Byte (Response)

Bit	Description
0	Reserved for future use
1	Reserved for future use
2	Reserved for future use
3	Reserved for future use
4	Error flag. 1 = command was unsuccessful. Data section of response packet contains the code of the error
5	Reserved for future use
6	Reserved for future use
7	Reserved for future use

TABLE 6.13

Error Codes

Code	Description
$01	Transponder not found
$02	Command not supported
$03	Packet BCC invalid
$04	Packet flags invalid for command
$05	General write failure
$06	Write failure because of locked blocks
$07	Transponder does not support command
$0F	Undefined error

for commands as well as their responses. Additionally, the protocol manages some aspects of the functionality of the reader and enhances the functionality of basic transponders by requesting the execution of compounded commands. This protocol is intended for point-to-point, half-duplex communications in which the host is the primary station. In some situations, however, the host may receive responses without sending a request first. This is a binary, byte-count-oriented protocol in which the data length is passed as a parameter in the message. Similar to the TIRIS protocol, the message from the host to the interrogator is called a *request*, while the message from the interrogator to the host is called a *response*.

The communication between host and interrogator is performed using a frame structure. The frame structure is essentially the same for requests and responses with the exception of the flag as they have different meanings. Requests use control flags, while responses use status flags. In this protocol, the length of the message data is limited to 4091 bytes. Table 6.14 describes the structure of the request and response frames.

In *pass-through-service* mode, the interrogator passes the information that it has received from the host to the transponder immediately without doing any other action. When the interrogator receives a response from the transponder, it also passes it directly to the host without further action. If the interrogator does not receive a response from the transponder in a specified period of time, it sends a message to the host with the error flag set and the

TABLE 6.14

Structure of Request and Response Frames

	Description	
Byte	**Request Frame**	**Response Frame**
1	Start Code[a]	Start Code
2	Data Length[b]	Data Length
3	Data Length	Data Length
4	Service Code[c]	Service Code
5	Control Flag	Status Flag
6	Message Data	Message Data
...	Message Data	Message Data
N-1	Message Data	Message Data
N+6	BCC	BCC
N+7	BCC	BCC

[a] The first byte contains the *start code*. This is the start frame delimiter using the value $D5.

[b] The next block is the two *data length* bytes that specify the number of data bytes that follow that block. The minimum length is 4 bytes (the service code, control flag, and two BCC bytes), while the maximum length is limited to 4095 bytes.

[c] The *service code* indicates the type of command that is being sent, as shown in Table 6.15.

TABLE 6.15

Service Codes

Code	Description
$01	Pass-through Service
$02	Single Service
$03	Compound Service
$04	Ancillary Service

appropriate error code. Using pass-through service, the host directly controls the transponder.

In *single-service* mode, after receiving a request from the host, the interrogator analyzes and verifies it before formatting and sending a single command to the transponder. Also, when the interrogator receives a response from the transponder, it analyzes it, verifies it, formats it, and sends a unique response to the host.

The *compound-service* mode is used by the host to instruct the interrogator to perform specific actions on transponders using the transponder protocol. The difference between these actions and the actions from a single-service request is that in compound service, the interrogator performs complex operations involving multiple commands before returning a response to the host computer. After receiving a compound-service request, the interrogator also analyzes and verifies it before sending the appropriate commands to the transponder. The interrogator also analyzes and verifies all the responses received from the transponder before sending a single response to the host. This response, however, may consist of more than one frame depending on its size. Compound service increases the performance of the system because, for example, it takes less time to send a single command to read four blocks of memory and transmit the information to the host, rather than issuing four different reading commands while transmitting the intermediate information to the host after each one has been performed.

Ancillary-service mode is used to instruct the interrogator to perform certain actions that do not involve communication with the transponder. Examples of such actions are interrogator reset, read interrogator version, and perform interrogator diagnostics. They are also used to upload new software versions in the interrogator or to set up new communication parameters.

Table 6.16 lists the available service codes with the exception of the pass-through-service codes as they are defined by the transponder protocol. The different transponder protocols are described in Chapter 7. The different commands are described in further detail in the next sections of this chapter. Table 6.16 can also be used to understand the type of commands available for each service code.

For control flags, four of the bits in byte 5 are used to specify instructions to the interrogator or they carry information about the message, as shown in Table 6.17.

TABLE 6.16

Commands Available for Different Service Codes

Command	Command Code	Single	Compound	Ancillary
		Service Code		
Tag Version	$03	$02		
Read Block	$12	$02		
Read Multiblock	$02	$02	$03	
Read Block SID	$FE		$03	
Read Multiblock SID	$FD		$03	
Write Block	$05	$02		
Write and Lock Block	$07	$02		
Write Multiblock	$06	$02	$03	
Lock Block	$08	$02		
Lock Multiblock	$09	$02	$03	
Repeat Last Request	$01			$04
Send Last Request	$02			$04
Stop Continuous	$04			$04
Start Synch	$05			$04
Reader Reset	$10			$04
Reader Version	$11			$04
Reader Diagnostic	$12			$04
Read Reader Setup	$13			$04
Start Flash Loader	$16			$04
Factory Lock Block	$3D	$02		
Write SID Code	$3E	$02		
Factory Programming Off	$3F	$02		

The next byte, byte 5, is the *control flags* byte. This byte has a different meaning depending on the direction of the message. For Request Frames (host-to-interrogator), byte 5 is called the *control flag byte*. For response frames (interrogator-to-host), byte 5 is called the *status flag byte*.

For status flags, four of the bits in byte 5 are used to provide information about the result of the request, as shown in Table 6.18.

Bit 0 (exception flag) indicates whether the request was successfully routed to the interrogator. When bit 0 is reset to 0, it indicates that no exception occurred and the request was completed satisfactorily. When bit 0 is set to 1, it indicates that an exception occurred. In this case, the error code that describes the exception can be found in the first byte of the message. These errors are shown in Table 6.19.

Bit 1 (more flag) informs the interrogator whether more data follow (bit 1 = 1). Bit 2 (emulation flag) informs the interrogator whether the transponder completed the request using a compound command to emulate a complex command (bit 2 = 1). Bit 3 (auto-repeat flag) informs the interrogator where

TABLE 6.17

Bits for Control Flags

Code	Description
0 (LSB)	Reserved
1[a]	More Flag
2[b]	Emulation Flag
3[c]	Auto Repeat Flag
4[d]	BCC Flag
5 to 7	Reserved

[a] When bit 1 (more flag) is set to 1, it indicates to the interrogator that it should expect more frames until the host sends a frame in which bit 1 is reset to 0. When bit 1 is reset to 0, it indicates to the interrogator that there are no more data associated with the request either in a single message or at the end of a series of messages.

[b] Bit 2 (emulation flag) indicates the interrogator if it is allowed to emulate a complex request by sending multiple single requests. This results in substituting a complex command with a compound command. When bit 2 is set to 1, this means that if the command cannot be performed using the current transponder version, the interrogator will attempt to use a compound command. When bit 2 is reset to 0, the substitution is not allowed.

[c] Bit 3 (auto-repeat flag) specifies whether or not the interrogator should automatically repeat the execution of the request. When Bit 3 is set to 1, auto-repeat is on. In this case, the interrogator will execute the request, send the response it has received from the transponder back to the interrogator, wait for a specified amount of time, and send the request again. This process is repeated until it receives another instruction with bit 3 reset to 0.

[d] Bit 4 is the BCC flag. It specifies the type of block code check used to ensure the integrity of the message. When bit 4 is reset to 0, the message uses CRC-CCIT, while when bit 4 is set to 1, the message uses LRC. Although LRC is simpler to implement and requires less processing time than CRC, it does not offer as much protection as CRC.

TABLE 6.18

Bits for Status Flags

Code	Description
0 (LSB)	Exception Flag
1	More Flag
2	Emulation Flag
3	Auto Repeat Flag
4	BCC Flag
5 to 7	Reserved

TABLE 6.19

Error Messages Associated to Exception Flag

Code	Description
$00	Reserved
$01	Request data corrupted and not executed
$02	Application not supported
$03	Data format error and request aborted
$04	Continuous mode not available for this request
$05	Reserved
$06	Reserved
$07	Reserved
$08	Reserved
$09	Reserved
$0A	Reserved
$0B	Reserved
$0C	Reserved
$0D	Reserved
$0E	Reserved
$0F	Undefined error and request aborted

TABLE 6.20

Structure of Single Service Messages

Length	Description
5 bytes	Frame
1 byte	Command
1 byte	Format
1 byte	Synch
≤ 4088 bytes	Message Data
2 bytes	Frame

the transponder performed the request once (bit 3 = 0) or continuously (bit 3 = 1). Bit 4 (BCC flag) specifies the method used to generate the BCC byte: CRC (bit 4 = 0) or LRC (bit 4 = 1).

The next byte or set of bytes are the *message data* bytes that carry the actual message. The actual format of the message depends on the type of command that was specified in the service code byte. In pass-through-service messages, the interrogator is transparent to the request from the host to the transponder. The communication from the host to the transponder uses the transponder protocol. Because this is a bit-oriented protocol, it may be necessary to add the necessary bits to the message to ensure that the whole frame falls within a byte-oriented protocol. These are called *pad bits*. The structure for single-service messages is shown in Table 6.20.

TABLE 6.21

Format Byte for a Request Message

Bit	Name	Description
0 (LSB)	Reserved	
1	Code Extension	0 = No extension code 1 = Code extension used
2	Addressing	0 = Not addressed 1 = Addressed. Address contained in first 4 bytes of data field
3	Format	0 = Fixed format 1 = Variable format
4	Transmitter control	0= Turn transmitter off after executing command 1 = Leave transmitter on
5	Reserved	
6	Reserved	
7 (MSB)	Reserved	

TABLE 6.22

Format Byte for a Response Message

Bit	Name	Description
0 (LSB)	Error	0 = No error 1 = Error occurred. Status code is first byte of data field after any address information
1	Code Extension	0 = No extension code 1 = Code extension used
2	Addressing	0 = Not addressed 1 = Addressed. Address contained in first 4 bytes of data field
3	Format	0 = Fixed format 1 = Variable format
4	Reserved	
5	Reserved	
6	Reserved	
7 (MSB)	Reserved	

The structure for single-service messages is shown in Table 6.20.

The format byte contains information about the format and status of the message. The contents are different depending on whether the message is a request or a response. Table 6.21 shows the description of the format byte for a request message, while Table 6.22 shows the description of the format byte for a response message.

Table 6.23 shows the meaning of the synchronization codes for both request and response messages.

The structure for compound messages is shown in Table 6.24.

Tables 6.25 and 6.26 show the structure of the format byte for request messages and response messages, respectively.

TABLE 6.23

Synch Byte for a Request or for a Response Message

Bit	Name	Description
0–2	Synchronization type	000 = No synchronization 001 = Master Slave synchronization (Master) 010 = Master Slave synchronization (Slave) 011 = Cascaded synchronization
3	Action on "No Transponder"	0 = All responses sent to host 1 = "No transponder" responses not sent to host
4	Alternate Operation during continuous mode	0 = All valid data responses sent to host 1 = Identical responses sent to host only once
5	Reserved	
6	Reserved	
7	Reserved	

TABLE 6.24

Structure of Compound Messages

Length	Description
5 bytes	Frame
1 byte	Command
1 byte	Format
1 byte	Synch
≤ 4088 bytes	Message Data
2 bytes	Frame

TABLE 6.25

Format Byte for a Request Compound Message

Bit	Name	Description
0 (LSB)	Reserved	
1	Reserved	
2	Addressing	0 = Not addressed 1 = Addressed. Address contained in first 4 bytes of data field
3	Format	0 = Fixed format 1 = Variable format
4	Transmitter control	0= Turn transmitter off after executing command 1 = Leave transmitter on
5	Reserved	
6	Reserved	
7 (MSB)	Reserved	

TABLE 6.26

Format Byte for a Response Compound Message

Bit	Name	Description
0 (LSB)	Error	0 = No error 1 = Error occurred. Status code is first byte of data field after any address information
1	Reserved	
2	Addressing	0 = Not addressed 1 = Addressed. Address contained in first 4 bytes of data field
3	Format	0 = Fixed format 1 = Variable format
4	Reserved	
5	Reserved	
6	Reserved	
7 (MSB)	Reserved	

TABLE 6.27

Structure of Ancillary Messages

Length	Description
5 bytes	Frame
1 byte	Command
≤ 4090 bytes	Message Data
2 bytes	Frame

Note: Because this type of command does not involve communication with the transponder, it does not require the format or sync bytes present in the other types of messages. The last two bytes contain the BCC code of either CRC or LRC. This is performed over all the preceding data with the exception of the start-of-frame byte.

The description of the synchronization bit is the same as for single messages that was shown in Table 6.23. The structure of ancillary messages is shown in Table 6.27.

The method of flow control demands for each request to receive a response. The flags in the response message indicate the presence of an error or other exceptions. After a certain time without receiving a response, the request can time out. The control is done by the host that acts as the primary station. Data lengths are variable for each message, although the maximum data length in a single message is 4091 bytes.

7

The Air Communication Link

CONTENTS

This chapter describes the communication methods used to transfer information from the interrogator to the transponder and from the transponder to the interrogator. The communication from interrogator to transponder is usually known as the *forward communications link*. This consists mostly of commands sent from the host computer to the transponder via the interrogator. It also contains the data to be written for those transponders whose memories can be rewritten. The communication from transponder to interrogator is known as the *return link* or the *backscattered communication*. This consists mostly of the transmission of the data stored in the transponder's memory.

While there are several standards that specify the critical elements of the communications protocol, the implementation of these standards especially varies not only among different manufacturers but also among different products from the same manufacturers. Therefore, this chapter does not pretend to be a comprehensive collection of all the air communication methods employed by RFID systems. Instead, this chapter explores the most common approaches to the communication between these elements of the RFID system using specific systems as general examples. This should give the readers the basic information while preparing them to search for more specific details if they so require.

This chapter starts by describing the different air communication protocols followed by the study of the elements that make up the forward and the return communications link. It continues by talking about the different modes of operation available to transponders. The last section in this chapter describes the different arbitration methods and the anticollision procedures that allow effective communication when there is more than a single transponder in the radiofrequency field.

7.1 Communication Protocols

7.1.1 Communication Protocols for Systems Operating in the LF and HF Ranges

The T5557 is a transponder with a 300-bit memory manufactured by Atmel® that operates in the low-frequency range and is similar to other transponders from different families of manufacturers. The communication between transponder and interrogators starts with the chip in the transponder being initialized after it has reached its threshold voltage. Afterward, the chip will be ready or will experience an additional delay of 67 ms depending on the status of one of its configuration bits. In read mode, the data from the memory are transmitted serially by modulating the load across the antenna terminals, starting with bit 1 from block 1. This process continues until the last bit from the specified number of blocks has been transmitted. The number of blocks to be read is defined by one of the bits in the transmitted command.

The transmission from the transponder to the interrogator uses amplitude-shift keying (ASK) modulation by switching a load connected to the antenna terminals on and off. Figure 7.1 shows the waveform across the load for different types of coding.

The transponder has the ability to insert a special pattern in order to synchronize the interrogator before transmitting the first block. This pattern consists of four bit periods in which the modulation is turned off and on, as shown in Figure 7.2.

In order to write data to the transponders, the interrogator uses on-off keying. It interrupts on and off the electromagnetic field generated as shown in Figure 7.3. The time between the gaps is used to encode the data to be transmitted: 24 periods of the RF field for a 0 and 54 periods of the RF field for a 1. The process finishes after the transponder has detected that there is no gap

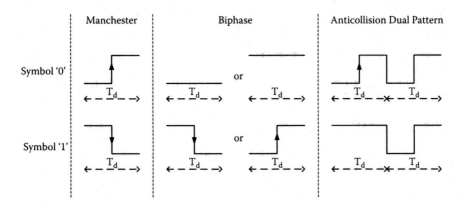

FIGURE 7.1
Waveforms across load for different types of coding.

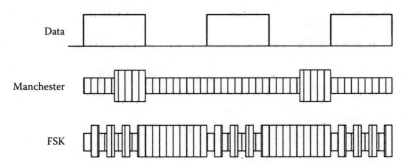

FIGURE 7.2
Synchronization patterns for Manchester and frequency-shift keying (FSK) modulations.

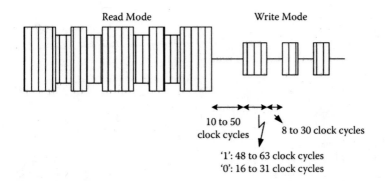

FIGURE 7.3
Gap sequences used in data transmission.

for more than 64 periods. The initial gap triggers the communication from the interrogator to the transponder. This initial gap can be accepted at any time after the system has been initialized.

After the initial gap, the transponder expects to receive a valid command sequence. A standard write command needs the operation code, the lock bit, the 32 bits that make up the data, and the 3-bit address. Writing to a protected transponder requires the additional 32-bit password that is placed between the operation code and the data address bits. The password bits are compared bit by bit with the contents of the password stored in memory. If the bits do not match, the transponder returns to read mode. The complete writing process is shown in Figure 7.4.

After all the information has been correctly received, the transponder can be programmed. There is, however, a delay of one clock period between the end of the writing sequence and the start of the programming. The typical programming time is 5.6 ms, including the reading for data verification. Figure 7.5 displays the voltage measured across the antenna after programming a memory block.

For the Tag-it™ family of transponders manufactured by Texas Instruments, the communication between interrogator and transponder is done by amplitude modulation of the carrier. This allows the interrogator to generate enough energy to power the chips on board the transponders while simultaneously sending data. The encoding for the logic symbols *1* and *0* is done

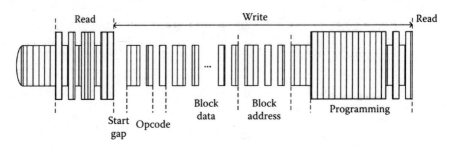

FIGURE 7.4
Writing sequence.

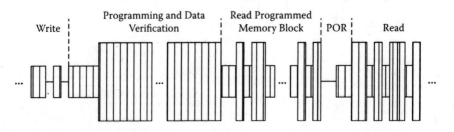

FIGURE 7.5
Voltage across antenna terminals after programming a block.

by pulse width modulation: *1* is encoded as a power burst lasting 150 µs, while *0* is encoded by power bursts lasting 100 µs. Data are framed between a start-of-frame (SOF) signal and end-of-frame (EOF) signal. SOF consists of a 50 µs burst followed by a standard duration pulse, terminating with a 150 µs pulse. The EOF signal consists of a 50 µs burst followed by a standard duration pulse, terminating with a 100 µs pulse.

The communication between the transponder and the interrogator uses the modulation of the antenna as it tunes and detunes the signal to generate the backscattered information. The signals being transmitted are encoded using a Manchester code, taking advantage of the clock frequency being embedded in the transmitted signal. With the load modulation, a Manchester high level is represented by a frequency of 423.75 kHz, while a Manchester low level is represented by a frequency of 484.29 kHz.

7.1.2 Communication Protocols for Systems Operating in the UHF Range

Transponders operating in the UHF range manufactured by Atmel® are based on the *interrogator talks first* (ITF) approach to establish communication with the interrogator. The communication always starts with a forward message, that is, a message from the interrogator to the transponder. After receiving this initial message, the transponder responds by backscattering a message to the interrogator. This is known as the *return link*. This backscatter communication is done using a binary phase-shift keying (2PSK) modulation based on changing the imaginary part of the input impedance in the transponder as a function of the message between sent.

The communication is based on a system of frames, as shown in Figure 7.6. The first frame is the header frame that is used to adjust the timing. The second frame is the data frame that is used to transport the data. The length of this frame depends on the command sent and the addressing mechanism being used. The last frame is the end-of-transmission (EOT) frame that marks the end of the transmission. Once the transmission has finished, the interrogator enters into listening mode after having received another header frame.

While the interrogator is transmitting data, the transponder remains in listening mode. During the transmission of the interrogator header frame, it is possible for the transponder to operate in full duplex. It transmits data back to the interrogator in a loop until the EOF frame is sent.

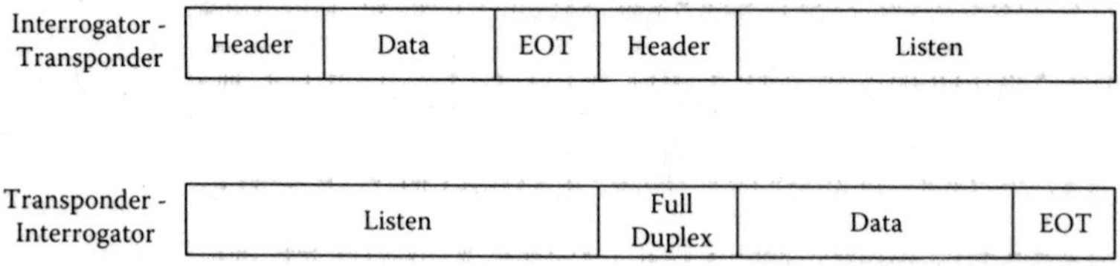

FIGURE 7.6
Frame communications during default transmissions.

The communication between transponder and interrogator must be reliable even in the noisy environments in which these systems normally operate. To improve the reliability of the communication link, the interrogator transmits periodic clock ticks to the transponders, ensuring reliability because it helps to block other radiofrequency sources. In addition, the link between transponder and interrogator operates synchronously, and because the interrogator controls the speed and modulation ticks, the resulting transmission has a higher signal-to-noise ratio than an asynchronous communication.

The general structure of the frames is different between the forward or return link as well as the type of commands that it carries. All the frames start with a header frame. The data frame in the forward link depends on the type of command carried, as shown in Figure 7.7, which shows frames for different types of commands. Data frames contain the command and its cyclic redundancy check (CRC), data-setting parameters, address information, and data for programming. For example, short commands consist of 6 bits and its 2-bit CRC, while long commands consist at least of the command fame, the CRC, and the parameter field. In the transmission of a long command, these frames are followed by a CRC frame of 16 bits. The transmission ends with an end-of-transmission frame that is made of two EOF symbols.

Figure 7.8 shows the general structure for the return link frames. The communication link starts with a header frame that is used to obtain reference symbols, adjust modulation, and set the timing for the EOF characters. This is followed by a data frame, a CRC frame 16 bits long, and an

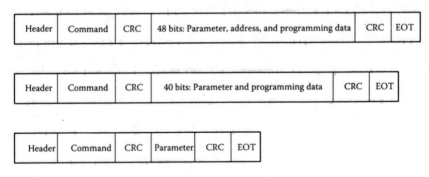

FIGURE 7.7
Examples of data frames in a forward link.

FIGURE 7.8
General structure for return link frames.

end-of-transmission frame. The data frame can have a length of 8 or 16 bits depending on the type of data carried. The silent frame shown in Figure 7.8 is only used during programming. During the silent frame period, the interrogator only transmits a continuous wave. The length of this period depends on the strength of the electromagnetic field.

The general structure of the communications flow for these transponders is shown in Figure 7.9. The transponder is continuously waiting for a notch indicating the presence of the external field. After it has been received, it will check for a valid header frame that indicates the beginning of a transmission. After the header frame has been received correctly, the transponder performs several additional verifications while transmitting the different parts of the message, including checking for the anticollision requests. The process ends after receiving the EOF data.

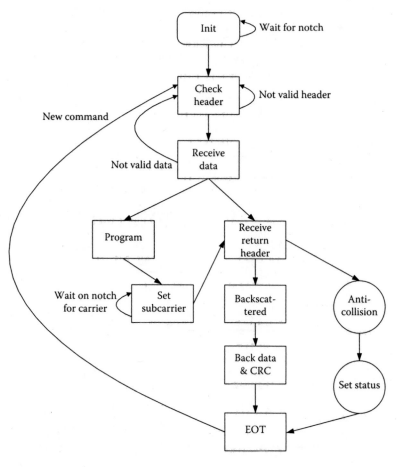

FIGURE 7.9
General structure of communications flow.

7.2 The Communications Link

7.2.1 Elements of the Communications Link

Because the transponder is powered by the electromagnetic field generated by the interrogator, it must measure the strength of this field continuously. If the strength of the field decreases in such a way that causes the voltage after the rectifier to be lower than a certain threshold, the circuits in the transponder activate an internal reset. For voltages higher than this threshold, the oscillator and the rest of the circuits in the transponder circuit are enabled. Figure 7.10 shows the time events between the different signals involved after the transponder is turned on for some transponders manufactured by Atmel®.

The duration of the power on reset (POR) period depends on how fast the signal ramps up. This in turn depends on the strength of the electromagnetic field. After the POR has executed, there is an additional waiting period called the *band gap* that is needed for stabilization of the rest of the internal circuits before the signal to enable the oscillator switches its state. Finally, there is a delay of 64 clock cycles (approximately 160 μs) before the oscillator in the transponder achieves operational status.

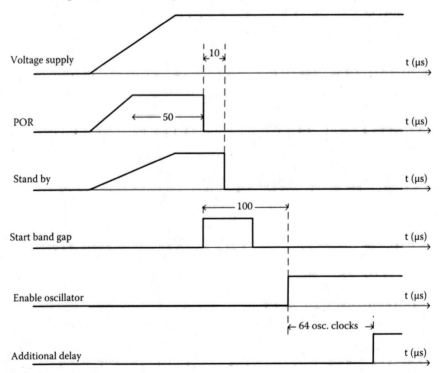

FIGURE 7.10
Chronogram of signals during power up.

When the electromagnetic field decreases, the voltage after the internal rectifier also decreases. Once it reaches a certain threshold level, the oscillator stops. This results in lowering the current required by the circuit that in turn results in the voltage supply decreasing with a lower slope as shown in Figure 7.11. Finally, if the voltage decreases below another threshold, the circuit executes a reset. The difference between the two voltage levels allows for taking into account variations in the strength of the field without having to execute a reset.

The communication between transponder and interrogator normally happens in a noisy environment. Therefore, the designers of the system must ensure bit error rates as low as possible. This is achieved by using a combination of several techniques including generating notches in the system clock, making the return link synchronous to notch signals generated by the interrogator, using different coding techniques, and using CRC to protect the integrity of the data. CRCs for commands are 2 bits long, while CRCs for data are 16 bits long. Figure 7.12 shows the concept of notch timing, which results from generating notches at periodic intervals in the signal generated. For a typical value of signal level between 450 mV and 600 mV, T_1 lasts 4 μs and T_{notch} lasts 2 μs.

Other techniques include using *pulse interval encoding* to define the symbols *1* and *0*, whose timing and timing differences are controlled by the interrogator as an additional element to increase the robustness of the link. The fact that the return link is synchronous to the interrogator allows it to close the communication as soon as it detects a transmission error.

The communication from the interrogator to the transponder is known as the *forward link*, while the communication from the transponder back to the interrogator is known as the *return link*. The forward link carries the different type of commands to the transponder, while the return link carries the data from the transponder to the interrogator. These data will only be available for those commands that require the return of data. Because the

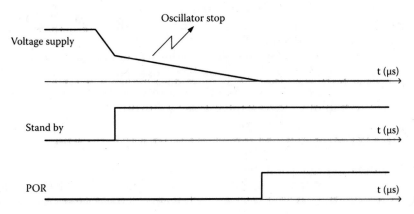

FIGURE 7.11
Chronogram of signals during power down.

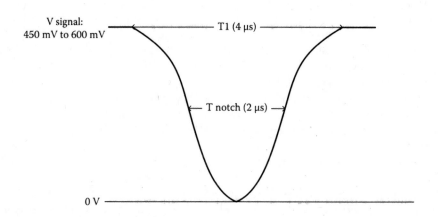

FIGURE 7.12
Notch timing chronogram.

handover mechanism from the forward to the return link happens synchronously, there is no need to wait for a cycle between them. Once the forward link sends its last frame that consists of an end-of-transmission message, the transponder is ready to transmit its return link header.

Texas Instruments RFID systems operating at low frequency use a frequency of 134.2 kHz to encode a low bit and a frequency of 123.2 kHz to encode a high bit. Because each bit uses 16 radiofrequency cycles to transmit, the duration of a low bit is different from the duration of a high bit. The low bit needs 119.9 μs to transmit, while a high bit needs 129.2 μs. The interrogator sends information to the transponder to transfer commands, addresses, or data. The writing process is always initiated by turning the radiofrequency transmitter on for a period of time of up to 50 ms. Afterward, the interrogator starts sending the bits that make up the command and associated data. Write bits last 2 ms. The differentiation between high and low bits is based on the duty cycle for each one of these bits. As shown in Figure 7.13, a high bit has a duty cycle approximately equal to 50%, while the low bit has a duty cycle approximately equal to 20%. The program phase consists of a continuous radiofrequency field for the duration of this period.

7.2.2 Forward Link

The forward header in most Atmel® transponders has four symbols, as shown in Figure 7.14. This header is accepted by the transponder if the length of the symbols *check1* and *EOF'* is larger than the length of the symbol *0'*.

Figure 7.15 shows that the information being transmitted is encoded in the timing between pulses. After the frame has been accepted and the

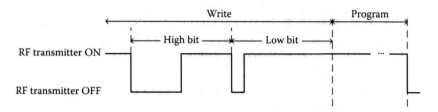

FIGURE 7.13
Write and program signals.

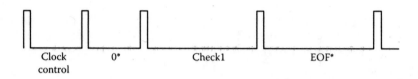

FIGURE 7.14
The four symbols in the forward header.

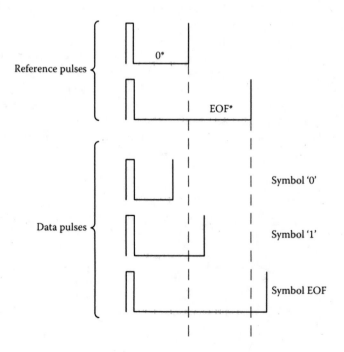

FIGURE 7.15
Pulse timing used to encode information.

interrogator transmits data, the transponder measures the time between two pulses or notch signals. This is used by the transponder to decide whether the received symbol is a *0* or a *1* by comparing its length with the length of the symbols *0*˙ and *EOF*˙, as shown in Figure 7.15.

This figure shows that a symbol will be interpreted as a *0* if its timing is shorter than the time that was sent in the Header for the symbol *0*˙. A symbol will be interpreted as a *1* if the time used by that symbol is longer than for the symbol *0*˙ but shorter than the time used in the header for the symbol *EOF*˙. Finally, EOF will be interpreted for a symbol that lasts longer than the *EOF*˙ symbol. This approach increases the reliability of the link as the transponder does not have to measure absolute but relative time intervals.

The communication from the interrogator to the transponder for the Series 2000 Texas Instruments RFID systems is based on transmitting bursts of energy of specific duration at specific intervals. For example, the commands used to read read-only or read/write transponders consist of an initial power burst lasting 50 ms followed by a period of 20 ms in which the radiofrequency field is turned off. In fact, all these transponders use the 50 ms power burst as the initial identifier for an upcoming command in addition to providing the energy to power the transponder. The following are some examples of formats for the different commands supported by these transponders. Figure 7.16 shows the protocol to transmit the command used to program a rewritable transponder.

The command starts with the 50 ms power burst followed by the specific command keyword and the password if necessary. These are followed by the 80 data bits used by this family of transponders and the 16 bits that make up the write frame. The last information being sent is the PBII, after which the transponder will read the new data stored in its memory to be verified by the interrogator. Figures 7.17 and 7.18 show the communication protocols for the commands *lock page* (of multipage transponders) and *selective read page* (of multipage transponders), respectively.

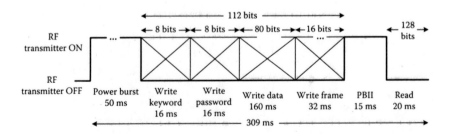

FIGURE 7.16
Protocol used to transmit a command to program a rewritable transponder.

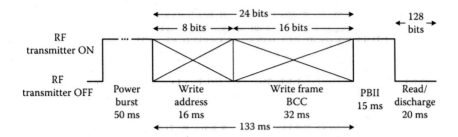

FIGURE 7.17
Protocol for the *lock page* command.

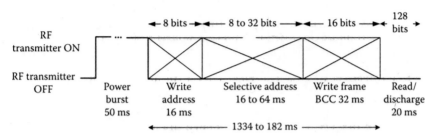

FIGURE 7.18
Protocol for the *selective read page* command.

7.2.3 Return Link

The return link in the majority of Atmel® transponders uses synchronous communication and angular modulation (binary PSK) in order to increase the signal-to-noise ratio and reduce the bit error rates because the return signals have extremely low power values. The synchronicity occurs as the transponder transmits its signal between two notches from the signal generated by the interrogator.

The return link starts with the return link header that is used to provide the timing information for the signal similar to the forward link header. Figure 7.19 shows the structure of the return link header for different types of coding styles used for the communication. In any case, the return link header always contains four symbols. The function and structure of each one of these symbols are shown in Table 7.1.

The return link header waits for a notch in the radiofrequency field generated by the interrogator before starting the generation of the reference timing. The data in the return link can be encoded using any of these coding techniques: bit stream (raw data), notch-locked non-return-to-zero-inverted (NRZI) encoding, soft-locked NRZI, FM0, and 3phase1. The EOT frame contains two EOF symbols.

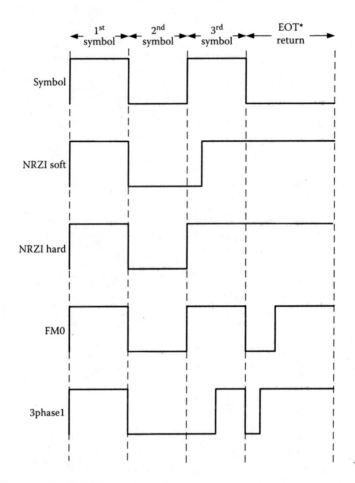

FIGURE 7.19
Structure of the return link header for different coding styles.

TABLE 7.1
Function of the Symbols in the Return Link Header

Symbol	Function	Additional Comments
1st	Main timing and power management	Used by interrogator to affect transmission rate of return link
2nd	Timing Reference	Used for anticollision control
3rd	Same timing as symbol 4	When using NIRZ, modulation is on
4th	Timing reference for EOF detection	

Figure 7.20 shows the transmission protocol used by the transponder to send information back to the interrogator as its response to a Read Data command for Texas Instruments transponders using their TIRIS protocol. All parts of this message are transmitted, with the least significant bit (LSB) being the first bit transmitted.

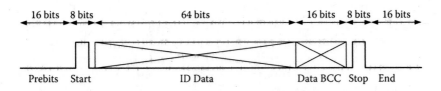

FIGURE 7.20
Transponder data sent back to the interrogator.

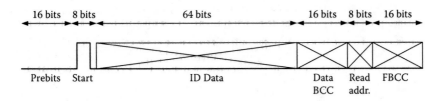

FIGURE 7.21
Return link for a multipage transponder.

After transmitting the pre-bits and eight bits that specify the start sequence, the transponder sends the 80 bits stored in its memory. The 64 first bits are the data bits, while the following 16 bits are the block check code (BCC) used to ensure the integrity of the data. These are followed by the stop and end bits. In order to be ready for a new activation, the transponder finishes the process by discharging its internal capacitor at the end of the transmission. The same protocol is used for rewritable transponders. Figure 7.21 shows the return link when accessing a multipage transponder.

In this case, following the BCC data, the transponder transmits the read address. This consists of a 2-bit status field that provides information about the function that that transponder has executed and a 6-bit page field that provides information about the page that was affected by the execution. The data format ends with reading the frame block check character (FBCC).

7.3 Transponder Operating Modes

7.3.1 Transponders Operating in the LF and HF Bands

After the transponders are initialized, they can enter into two different states, *interrogator talks first* (ITF) or *public mode* or *electronic surveillance*, depending on the status of their operating mode flags. In ITF mode, the data stream received through modulation from the interrogator undergoes different verifications. If they all result in a valid status, the transponder is synchronized with the data stream. At that point, the transponder is ready for communication. This communication always starts with a forward message, which is a message from the interrogator to the transponder. This communication method is used in multitransponder applications due to its anticollision capabilities. In this mode of operation, the transponder can be in any of the following states: *power down, ready, selected,* or. *quiet.*

When the transponder is not able to harvest enough energy from the surrounding electromagnetic field to energize its internal chip, it enters power down state. The transponder can also enter the ready state after having received a reset command. The transponder enters the selected mode after having received an explicit command with its own serial number from the interrogator. It is possible to set several transponders into the ready state by sending a specific command containing a matching partial identification number. It is necessary to set a transponder into the selected mode before being able to access it. The quit state is entered after a selected state after the interrogator selects another transponder or receives a specific command to enter this state.

Public mode communications start with the SOF pattern, which is followed by continuously streaming the data from the user memory. As shown in Figure 7.22, the first bit being sent is bit 31 from data block 0. When the transponder reaches its final data block, it starts transmitting data block 0 again. This process terminates when the electromagnetic field that energizes the transponder is turned off or when the transponder receives the appropriate command from the interrogator.

Electronic surveillance mode is a particular operating mode in which the transponders are attached to specific retail articles. By programming all bits in block 0 of the user memory to 0, the transponder is set to an *unpaid state.* After the article has been purchased, the cash register sets the bits of block 0 to 1, thus entering the *paid state.* Additional security can be achieved by protecting the bits of block 0 with a write password. If a transponder in the *unpaid state* enters the electromagnetic field generated by the interrogator located at the doors of the store, it will modulate the field. Consequently, the presence of a transponder marked in the *unpaid state* will be detected by the

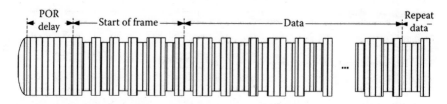

FIGURE 7.22
Initial sequence for the public mode.

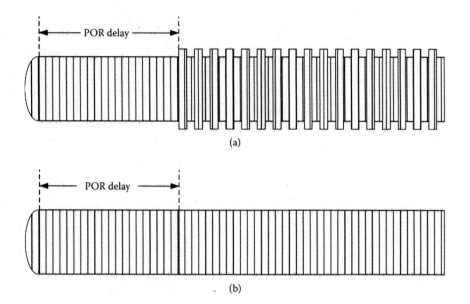

FIGURE 7.23
Signals from the transponder in electronic surveillance mode: (a) article unpaid, and (b) article paid.

interrogator, triggering the appropriate alarms. Conversely, a transponder marked as being in the *paid state* will not alter the electromagnetic field and will be transparent to the interrogator. Figure 7.23 shows the signals for these two cases.

Figure 7.24 shows the relationships between the different states for the transponders depending on their operational mode.

7.3.2 Transponders Operating in the UHF Bands

The possible states for the transponders manufactured by Atmel® operating in the UHF band are RF Field OFF, RF Field ON, Start, Mute, Observe, Ready, Active, and Isolated. These states and the paths between them are shown in Figure 7.25 and discussed here.

- The initial state is the *RF Field Off* state. Here, the transponder is not able to operate because there is no external radiofrequency field for which it can draw power. The transponder is able to store some configuration information for about 8 seconds after the field has ceased. After this time, the contents of these registers become undefined.

- When the transponder enters a region with an existing radiofrequency field whose magnitude allows the transponder to power

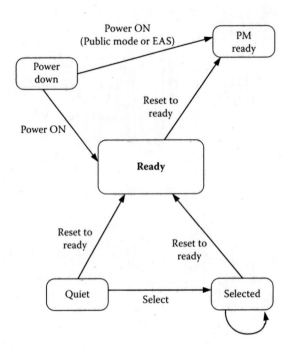

FIGURE 7.24
Different transponder states.

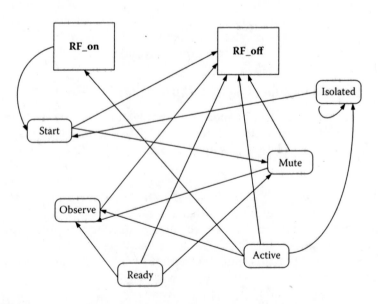

FIGURE 7.25
Operational states for UHF transponders.

its internal circuitry, this generates a power on reset. This is the *RF Field On* state. After the POR has occurred, the internal oscillator in the transponder is switched on. It is important to note that for this process to occur, the frequency of the external radiofrequency field must match the resonant frequency of the transponder's antenna and associated circuitry. Once this process has finished, the transponder moves into the *Start* state.

- In the *Start* state, the transponder refreshes the contents of the status registers and reads their configuration from its memory, enabling or disabling its trigger function. After this process, the transponder is able to detect the modulation ticks that signify information being sent from the interrogator. From this state, the transponder can move to the *Mute* state or to the *Observe* state.

- In the *Mute* state, the transponder cannot respond to operations such as programming, arbitration, or backscattering data. The transponder leaves the *Mute* state after some of its internal configuration registers have been modified.

- The *Observe* state allows the transponder to become synchronous with the interrogator for performing operations such as programming. To become synchronous the transponder checks that the forward header received from the interrogator is valid, that the command received is within the list of known commands, and that its CRC is also valid. If all these checks are passed successfully, the transponder moves to the *Ready* state.

- During the *Ready* state, the transponder receives the parameters associated with the type of commands sent by the interrogator. If any of the parameters are false, the transponder enters the *Mute* state. Otherwise, the transponder enters the *Active* state.

- In the *Active* state, the transponder is able to execute the command sent by the interrogator after receiving and checking the rest of the forward frame. If there are any errors in the information sent after the header frame, the transponder moves into an *Isolated* state or *Mute* state depending on the type of errors received. If the information sent in the forward frame is correct, the transponder executes the command. Once the transponder has finished the execution it moves back into the *Ready* state. Figure 7.26 shows with more detail the processes that occur during the *Active* state.

- The first step in this process is to verify the length of the received stream. If this is correct, the transponder checks for the validity of the position of the end-of-transmission frame. Depending on the result of this check, the transponder knows if the anticollision procedure has been activated in order to select that specific transponder. If this is the case, the transponder will enter the appropriate

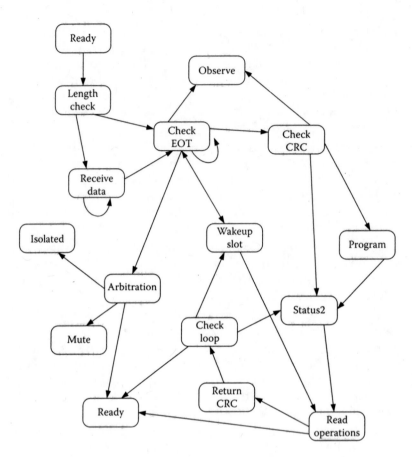

FIGURE 7.26
Processes during the active state of transponders.

anticollision and arbitration procedures. The transponder can also enter into receiving wakeup commands or other types of commands sent by the interrogator.

- In the *Isolated* state, the transponder is isolated and waits for either a POR or reset command to leave this state.

7.4 Arbitration for Transponders

7.4.1 Principles of Arbitration

When there is more than one transponder immersed in the electromagnetic field generated by an interrogator, it is necessary to develop and establish a procedure to select a single transponder or a specific group of transponders for carrying out specific functions. Alternatively, it can also be necessary to isolate a specific transponder or a group of transponders so they do not disturb communications. In some applications it may also be interesting to know whether there are transponders in the field or not. All these functions require the use of arbitration procedures. The two main types of arbitration procedures are *deterministic arbitration procedures* and time-slot based procedures, also known as *aloha procedures*.

Deterministic procedures are based on a *tree-walking function* in which the interrogator acting as the control node uses the individual IDs of transponders to select one or a group of them. The interrogator starts by polling all the transponders in the field. If there is more than one, their responses will collide and the interrogator will not receive the message. Therefore, in the next step, the interrogator polls only a specific subgroup of transponders based on their IDs. If the responses also collide, the interrogator reduces the population by polling only a specific subgroup from the previous subgroup. This process is repeated until the interrogator receives a response from a single transponder.

Aloha procedures are controlled by slot commands. After the transponders are turned on, they calculate random values for as long as they are powered. Once the transponder has received a slot command, it creates a random number from its generator, which points to a slot number in the chip. After receiving a start command, the transponders transmit their information to the reader during the first slop. If the random number matches the first slot number, it calculates a new one during this slot. When the random number matches the slot number controlled by the interrogator, that transponder becomes active. This allows the transponder to communicate with the interrogator. If the numbers do not match, the transponder enters a mute state and therefore does not respond to any commands. When the transponder receives a new slot command, it calculates if it is supposed to be active or mute.

7.4.2 Principles of Anticollision

The aim of anticollision procedures is to detect and identify the transponders that are present within the range of the electromagnetic field generated by the interrogator. While the implementation of specific anticollision procedures may vary between manufacturer and models, the principles of anticollision are common to all of them. For transponders manufactured by Atmel®, the communication with a single transponder or a group of transponders is controlled by the interrogator. The arbitration is initiated by the interrogator issuing a command for the transponders to transmit their identification numbers. The command can be tailored to the transponders that the interrogator expects to be in its area of influence. If the interrogator does not know anything about the transponders that are expected to be, the command is general and the detection encompasses all transponders. Otherwise, the command can use parameters such as a partial transponder identification number that will cause only transponders with partially matching IDs to respond to the command.

After the interrogator receives responses from the transponders, it scans their ID number from the most significant to the least significant bits. Each time slot corresponds to a specific bit position in the ID number. Because all transponders respond simultaneously with their modulated signal, their responses will be superimposed on one another. Therefore, because of the modulation being used, a damped signal overwrites a nondamped signal. With this, a logical 1 prevails over a logical 0. Consequently, the transponders with these ID numbers are removed from the process. When the interrogator detects a logical 1, it responds by broadcasting a gap in the radiofrequency field. The detection of a logical 0 does not originate any response from the interrogator. The gap in the field that is created as a response to the logical 0 is monitored by the transponders in the field and deduces, by comparing their ID number to the gap that was just broadcasted, if that specific transponder should continue in the current anticollision loop or eliminate itself from it. The transponders that are eliminated are muted and return to the ready state, in which they will remain until the interrogator initiates a new anticollision loop.

The transponders that have not been eliminated remain in the current loop as long as the final bit of the identification number has not been reached yet, and the process repeats itself. By the time the last bit has been interrogated, there will be a single transponder active as all the other transponders will have been eliminated through the previous steps. At this point, the interrogator is able to identify the associated ID for that specific transponder, and the transponder enters into the Selected state. Now, the transponder can establish direct communication with the interrogator and can be read or written. The transponder remains in the Selected state until the interrogator starts a new anticollision loop.

The MCRF450 family of transponders manufactured by Microchip Technology also supports anticollision. Their algorithm is based on time division multiplexing the transponder responses in which each transponder is only allowed to communicate with the interrogator during its time slot. This algorithm contains four control loops: Detection, Processing, Sleeping, and Reactivation.

The control of these loops is done by five commands as well as the transponder internal flag bits. The commands that control the anticollision loops are as follows:

- *Fast read request (FRR)*. With an internal flag set, the transponder will respond only to an FRR command. An FFR command consists of five timed gap pulses, as shown in Figure 7.27, in which the period of the gap contains specific information. There are a total of nine possible pulse positions allowed in this protocol.

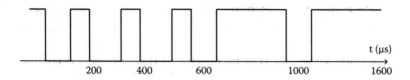

FIGURE 7.27
Structure of a fast read request (FRR).

- *Fast read bypass (FRB)*. This command is used in the reactivation loop. Upon reception of this command, the transponder responds with a block of data. The structure of this command is similar to the FRR command, with the difference being in the period of the gaps.
- *Matching code 1 (MC1)*. This command is used with the transponder and does not require any further processing. The command consists of time calibration pulses and pulse position modulation (PPM) signals. Upon the reception of this command, a transponder in the detection loops enters the sleeping loop.
- *Matching code 2 (MC2)*. When a transponder in the detection loop receives this command, it enters the processing loop. The structure of this command is similar to that of MC1.
- *End process (EP)*. This command causes a transponder in the processing loop to exit it and enter the sleeping loop.

Transponders can enter the detection loop by waiting for an FFR command or by transmitting a fast read (FR) response without waiting for a command. The first operation mode is called *interrogator talks first* (ITF), while the second operation mode is called *tag talks first* (TTF). The read and write operations are carried out during the processing loop. Only transponders with a special flag set in this loop respond to write or read operations. The sleeping loop is used to keep all processed devices in a silent condition in which they do not respond to external commands. The reactivation loop is used as an intermediate step by the interrogator to set transponders into different loops depending on the command that is transmitted to transponders in this loop.

7.4.3 Deterministic Anticollision Procedures

Deterministic anticollision procedures are based on the transponder backscattering the information to the interrogator only if the result of a compare condition is true. The compare condition is set up by the interrogator while broadcasting a specific command. Deterministic anticollision procedures are also called tree-based procedures. The interrogator sends a series of queries, each one with a prefix that is part of the transponder's ID code. Only those transponders with matching IDs respond to the interrogator's query. If there is more than one transponder with matching partial IDs, these transponders respond at the same time and their messages collide. The interrogator recognizes the collision of messages as the existence of more than one transponder with the prefix that was sent. At this point, the interrogator further refines its prefix, making it one bit longer, and the process repeats itself. Because all the transponders have different IDs, there will be a point at which only one transponder will respond to the query. This uniquely identifies the transponder that responds to the specific prefix.

Example 7.1:

Assume three transponders are in the read region of an interrogator with the following IDs. Explain the deterministic anticollision process.

        Transponder 1: 1111
        Transponder 2: 1101
        Transponder 3: 0010

Solution:

Let's assume that the interrogator starts its process by broadcasting the prefix *0*. Upon receiving this query, only transponder 3 responds because the first bits of transponders 1 and 2 do not match the prefix broadcasted. Because the interrogator receives a message from a transponder that has not collided with messages from other transponders, it uniquely identifies transponder 3 as being the only transponder in the field with an ID that starts with *0*.

Let's assume now that the interrogator starts its process by broadcasting the prefix *1*. Upon receiving this query, transponder 1 and transponder 2 respond, and therefore their responses collide. The interrogator then knows that there is more than a single transponder with an ID number starting with *1*. The interrogator proceeds to refine its query by increasing the original prefix by one bit, broadcasting the new prefix as *10*. Upon receiving this query, none of the transponders responds. This indicates to the interrogator that none of the transponders that had collided with the message *1* collides now. Therefore, the IDs of the transponders must contain the prefix *11*. The interrogator continues by increasing the length of the prefix, now to *110*. After receiving this query, only transponder 2 responds, and thus is uniquely identified.

This process does not require transponders with memory other than the memory used to store their ID numbers. This makes it an attractive protocol when used with low-cost transponders.

7.4.4 Aloha-Based Anticollision Procedures

Aloha-based procedures are based on a slot mechanism. In its most basic form, this mechanism starts working when the transponder calculates a slot out of 32 possible slots based on a random number. Within each slot, a transponder can communicate with the interrogator only if the calculated slot value is the same as the slot number that is under the interrogator's control. Figure 7.28 depicts the algorithm used in the most basic aloha protocol for anticollision.

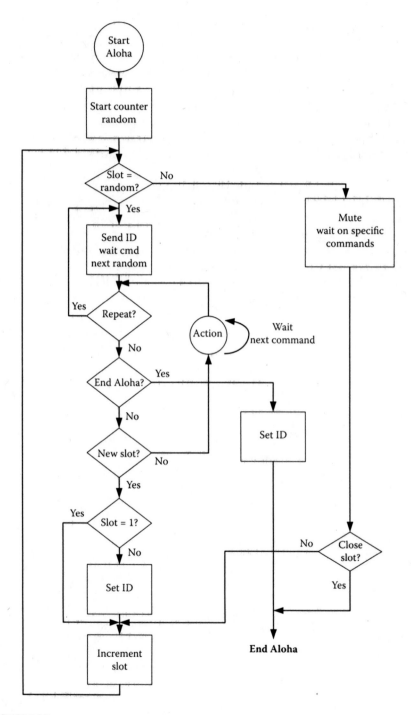

FIGURE 7.28
Structure of the basic aloha algorithm.

After a transponder receives a slot command, it broadcasts the random value as well as the contents of its identification number. The identification number is also broadcasted if the slot number is the same as the number transmitted by the reader and the transponder was selected previously. The slot is then incremented by one repeating the process.

Aloha procedures can be further divided into three versions: slotted aloha, framed slotted aloha, and dynamic framed slotted aloha. Slotted aloha follows the procedure described earlier. That is, the time is divided into several slots and transponders can only transmit data during the time slot they have selected. Framed slotted aloha is a variation of the basic slotted aloha. Here, one frame consists of several slots and the transponder chooses one slot in the frame to transmit data. The interrogator decides the size of the frame. This size remains constant during the process. Transponders generate a random number that is used to select a slot in a frame. Because the interrogator identifies transponders with multiple frames, it can identify the diverse transponders. The process for dynamic framed slotted aloha is similar, although in this variant the size of the frame is variable. This results in a more efficient algorithm for identifying transponders.

8

Commands for Transponders

CONTENTS

The communication from interrogators to transponders consists mostly in the interrogators issuing commands to be executed by the transponders. After the transponders have received the commands, they transmit information back to the interrogator. This information can be the requested data, status of their internal parameters, acknowledgment flags, or error codes.

This chapter explores the structure and configuration of commands for transponders as well as the different status and error messages transmitted by the transponders as response to these commands. Because of the diversity in the number of transponders in the market, this chapter focuses on describing the commands in representative examples of transponders. This will give the reader enough information to understand the structure of commands used in other types of transponders. This chapter starts by describing the commands in the Series 2000 systems from Texas Instruments that operate in the low-frequency (LF) band. It continues by studying the commands used in diverse transponders from Texas Instruments operating in the high-frequency (HF) band. The third section describes the commands supported in the Series 4000 interrogators from Texas Instruments that can operate in the LF and HF bands. The next section describes the commands used by the transponders from Texas Instruments operating in the ultra-high-frequency (UHF) band.

Finally, the last section in this chapter is focused on the commands used by transponders manufactured by Atmel® in the different frequency ranges.

8.1 Commands in the Texas Instruments Series 2000 Systems

8.1.1 Commands Supported by the TIRIS Bus Protocol

The commands supported by the TIRIS Bus Protocol can be divided into four different types of commands:

> *Group 0*: These commands are generally used to control the queue of the message traffic. They have codes $00 to $1F.
>
> *Group 1*: These commands typically result in reading or writing the RFID transponders. They have codes $20 to $3F.
>
> *Group 2*: These commands set up parameters in registers that execute auxiliary input/output functions. They have codes $40 to $5F.
>
> *Group 3*: These commands are for user add-on tasks. They have codes $60 to $7F.

Table 8.1 describes the commands in Group 0. It is necessary to note that although the protocol can support 32 different commands, there are only five commands actually implemented. All these commands operate as *immediate commands*.

TABLE 8.1

TIRIS Bus protocol Group 0 commands

Command Number	Description
$00[a]	Send count of records
$01[b]	Send next record from queue
$02[c]	Send record N from queue
$03[d]	Resend last record
$04[e]	Clear queue
$05	Reserved
...	...
$1F	*Reserved*

[a] Command $00 (*Send count of records*) returns the number of records that are buffered in the queue. This command does not require one to pass any additional data or parameters. The response is stored as an unsigned character.

[b] Command $01 (*Send next record from queue*) returns the next message that is buffered in the queue. This command does not require data either.

[c] Command $02 (*Send record N from queue*) returns the Nth message from the queue. This command requires the value N as its data argument. The response is stored in a record.

[d] Command $03 (*Resend last record*) resends the last sent message. This command does not require any data, nor does it produce data as response.

[e] Command $04 (*Clear queue*) clears all the messages buffered in the queue. The command does not require any data, nor does it produce data as response.

TABLE 8.2

TIRIS Bus Protocol Group 1 Commands

Command Number	Description
$20[a]	Charge only read
$21[b]	Read page N
$22[c]	Read page N (80 bit)
$23[d]	Selective read
$24	Reserved
$25	Reserved
$26	Reserved
$27	Reserved
$28	Reserved
$29	Reserved
$2A	Reserved
$2B[e]	Program 64 bit
$2C[f]	Program page N
$2D[g]	Program page N (80 bit)
$2E[h]	Selective program
$2F[i]	Program 80 bit to read/write transponder
$30	Reserved
$31	Reserved
$32[j]	Lock page N
$33[k]	Lock page N (SAMPT)
$34	Reserved
…	…
$3F	Reserved

[a] Command $20 (*Charge-only read*) performs a single charge-only read. Its execution does not require any parameters. The response contains a status code and a data field whose length is variable depending on the status code. The meaning of the different status codes is shown in Table 8.3. A data field length equal to 1 corresponds to one of the following status codes:
NO_READ
INCOMPLETE
MPTRERR_SPC_DATA
MPTRERR_STATUS
   A data field length equal to 9 corresponds to the following status codes, with the data fields 1 to 8 containing the ID of the transponder being read:
RO_TRP
RW_TRP
MPTCOTRP_U
MPTCOTRP_L
   The meaning of the different status codes is shown in Table 8.3.

[b] Command $21 (*Read page N*) performs a single read of the specified page (*N* page) in a multi-page transponder. This command needs to have the number of the page to read as a parameter. This parameter can range from $01 to $3F. The response contains a data field of variable length depending on the different status codes. A data field length equal to 1 corresponds to the same status codes as Command $20. A data field length equal to 9 corresponds to the following status codes, with the data fields 1 to 8 containing the ID of the transponder being read:

(*Continued*)

TABLE 8.2 (CONTINUED)

TIRIS Bus Protocol Group 1 Commands

RO_TRP
RW_TRP
   A data field length equal to 10 corresponds to the following status codes. Here, data fields 1 to 8 contain the ID of the transponder and data field 9 contains the read page number:
MPTRP_U
MPTRP_L
MPTRERR_PAGE_U
MPTRERR_PAGE_L

[c] Command $22 (*Read Page N [80 bit]*) performs a single read of the specified page in a multi-page transponder with 80 bits. It needs a parameter that specifies the page to be read. The response contains a data field of variable length depending on the different status codes. A data field length equal to 1 corresponds to the same status codes as Command $20. A data field length equal to 11 corresponds to the following status codes, with the data fields 1 to 10 containing the ID of the transponder being read:
RO_TRP
RW_TRP
RO_TRP_80
RW_TRP_80
   A data field equal to 11 corresponds to the following status codes. Data fields 1 to 10 contain the ID of the transponder being read, while data field 11 contains the read page number:
MPTRP_U
MPTRP_L
MPTRERR_PAGE_U
MPTRERR_PAGE_L
MPTRP_80_U
MPTRP_80_L

[d] Command $23 (*Selective read*) performs a selective read of a specified page in a multipage transponder. It needs the following three parameters: the type of transponder to read, the page to read, and the address within the page. The response contains a data field of variable length depending on the different status codes. A data field length equal to 1 corresponds to the same status codes as Command $20. A data field length equal to 9 corresponds to the following status codes, with the data fields 1 to 8 containing the ID of the transponder being read:
RO_TRP
RW_TRP
   A data field length equal to 10 corresponds to the following codes, with data fields 1 to 8 containing the ID of the transponder being read and data field 9 containing the number of the page:
MPTRP_U
MPTRP_L
MPTRERR_PAGE_U
MPTRERR_PAGE_L

[e] Command $2B (*Program 64-bit transponder*) writes a 64-bit ID into a read/write transponder. It needs a parameter with the information to write. The response always has a data field equal to 1 with the following possible status codes:
PROG_OK
PERR_FALSE_ID
NO_READ
INCOMPLETE
RO_TRP

TABLE 8.2 (CONTINUED)

TIRIS Bus Protocol Group 1 Commands

ᶠ Command $2C (*Program page* N) performs a write in the Nth page of a multipage transponder. It needs two parameters: the page to write and the 64-bit data to write in the transponder. The response always has a data field equal to 1 with the following possible status codes:

PROG_OK
PERR_FALSE_ID
NO_READ
INCOMPLETE
MPTPRERR_STATUS
MPTPERR_UNREL
MPTERR_PAGE_U
MPTERR_PAGE_L
MPTPERR_SPC_DATA
MPTPERR_LOW_VOLT
MPTPERR_LOCK
RO_TRP
RW_TRP

ᵍ Command $2D (*Program page N [80 bit]*) performs a write in the Nth page of a multipage transponder. It needs two parameters: the page to write and the 64-bit data to write in the transponder. The response always has a data field equal to 1 with the same status codes as Command $2C.

ʰ Command $2E (*Selective programming*) performs a selective write into the Nth page of a multipage transponder. It needs four parameters: the type of multipage transponder, the page number to write to, the address, and the ID of the transponder. The response always has a data field equal to 1 with the same status codes as Command $2C.

ⁱ Command $2F (*Program 80-bit data to transponder*) writes 80-bit data into a read/write transponder. It requires the 80-bit data to write as parameter. The response always has a data field length equal to 1 with the same status codes as for Command $2B.

ʲ Command $32 (*Lock Page* N) locks the specified page in a multipage transponder. It requires the page number as a parameter. The response always has a data field length equal to 1 with the following status codes:

MPTLERR_FS_DROP
MPTLERR_UNREL
MPTLERR_STATUS
MPTLERR_PAGE_U
MPTLERR_PAGE_L
RO_TRP
RO_TRP

ᵏ Command $33 (*Selective lock page* N) performs a selective lock of the specified page in a multipage transponder. It requires the following three parameters: the type of multipage transponder, the page to be locked, and the address. The response always has a data field length equal to 1 with the same status codes as Command $32.

TABLE 8.3

Status Code Name and Description for Results of Group 1 Commands

Status Code Value	Status Code Name	Status Code Description
$00	RO_TRP	Read Only transponder successfully read
$01	RW_TRP	Read/Write transponder successfully read
$02	MPTCOTRP_U	Successful read of unlocked page 1 in a Multipage transponder as a result of charge-only read
$03	MPTCOTRP_L	Successful read of locked page 1 in a Multipage transponder as a result of charge-only read
$04	MPTRP_U	Successful read of an unlocked page of a Multipage transponder as a result of a read command
$05	MPTRP_L	Successful read of a locked page of a Multipage transponder as a result of a read command
$06	MPTRP_80_U	Successful read of an unlocked page of Multipage transponder as a result of a read command in 80 bit mode
$07	MPTRP_80_L	Successful read of a locked page of Multipage transponder as a result of a read command in 80 bit mode
$08	RO_TRP_80	Read Only transponder successfully read in 80 bit mode
$09	RW_TRP_80	Read/Write transponder successfully read in 80 bit mode
$30	PROG_OK	Writing to a Read/Write or Multipage transponder completed successfully
$31	LOCK_OK	Locking of a Multipage transponder completed successfully
$40	NO_READ	No transponder data received
$41	INCOMPLETE	The start byte of the transponder was detected, but CRC failed
$42	MPTERR_CRC	Frame of multipage transponder is correct, but the data CRC failed
$43	MPTRERR_ STATUS	Returned invalid status after reading a Multipage transponder
$44	MPTRERR_ PAGE_U	Read unlocked of a multipage transponder but it is different than the requested page
$45	MPTRERR_ PAGE_L	Read locked of a multipage transponder but it is different than the requested page
$46	MPTRERR_ SPC_DATA	Special data status returned reading a multipage transponder
$47	MPTRERR_ STATUS	Invalid status returned writing a multipage transponder
$48	PERR_FALSE_ IDE	Different ID returned from writing a Read/Write or a Multipage transponder
$49	MPTRERR_ LOW_VOLT	Programming voltage is too low to write a multipage transponder
$4A	MPTRERR_ UNREL	Writing to a multipage transponder not reliable
$4B	MPTRERR_ LOCK	Attempt to write a locked page of a multipage transponder
$4C	MPTRERR_ SPC_DATA	Received special data status on writing a multipage transponder

TABLE 8.3 (CONTINUED)

Status Code Name and Description for Results of Group 1 Commands

Status Code Value	Status Code Name	Status Code Description
$4D	MPTRERR_ PAGE_U	Read unlocked page of a multipage transponder as a result of writing but it is different than the requested page
$4E	MPTRERR_ PAGE_L	Read locked page of a Multipage transponder as a result of writing but it is different than the requested page
$4F	MPTRERR_ STATUS	Invalid status returned on locking a multipage transponder
$50	MPTRERR_FS_ DROP	Field strength dropped whole locking a multipage transponder
$51	MPTRERR_ UNREL	Locking of a multipage transponder not reliable
$52	MPTRERR_ PAGE_U	Read unlocked page of a multipage transponder as a result of writing but it is different than the requested page
$53	MPTRERR_ PAGE_L	Read locked of a multipage transponder as a result of writing but it is different than the requested page

All Group 1 commands can be sent requiring an immediate, broadcasted, or queued response (see Table 8.3).

Table 8.4 shows the Group 2 commands. As with Group 0 and Group 1 commands, several command codes are reserved for future implementation.

Group 3 commands define user add-on tasks. For this reason, each use may develop his or her specific commands according to the intended applications.

8.1.2 Commands Supported by the ASCII Bus Protocol

The command supported by the ASCII Bus Protocol is divided into three types: commands for use with 64-bit read/write transponders, commands for use with multipage transponders, and general commands that can be used with either group of transponders. Table 8.5 lists the general commands for the ASCII Bus Protocol.

Example 8.1:

Describe the communications from the host computer to the interrogator using Command H.

Solution:

Host computer → Interrogator	H
Interrogator → Host computer	H (echoes it back)
Host computer → Interrogator	I2 (additional parameters)
Interrogator → Host computer	03<CR><LF>
	Return ports status: 0 to 3
	I/O lines 0&1 port 0...3 set to 1
	I/O lines 2&3 port 0...3 set to 0

TABLE 8.4

TIRIS Bus Protocol Group 2 Commands

Command Number	Description
$40[a]	Get Version
$41[b]	Set RF parameter
$42[c]	Get RF parameter
$43[d]	Set Antenna
$44[e]	Get Antenna
$45[f]	Write *electrically erasable programmable read-only memory* (EEPROM)
$46[g]	Read EEPROM
$47	Reserved
$48	Reserved
$49[h]	Read byte from input
$4A	Reserved
$4B	Reserved
$4C	Reserved
$4D[i]	Write byte to output
$4E	Reserved
$4F	Reserved
$50	Reserved
$51	Reserved
$52	Reserved
$53	Reserved
$54	Reserved
$55[j]	Get count of buffered ID records
$56[k]	Set Read Mode
$57[l]	Delete ID records
$58[m]	Get Read Mode
$59	Reserved
$5A[n]	Get ID records from ID buffer
$5B[o]	Put ID records to the ID buffer
$5C	Reserved
$5D[p]	Read block of data from memory
$5E[q]	Write block of data to memory
$5F[r]	Reset Interrogator

[a] Command $40 (*Get version*) returns a string of characters that contain the software version for the interrogator. This is a command with an immediate response that does not require any parameter.

[b] Command $41 (*Set RF parameter*) sets the period of time used to energize the transponder (charge time) and the duty cycle pause. This command can generate an immediate or a broadcast response. The command requires two parameters: the charge period and the duty cycle pause.

[c] Command $42 (*Get RF parameter*) returns the current RF setting as for the current charge time and duty cycle pause. This command can also generate an immediate or broadcast response. The command does not require any parameters. The response contains two fields: the charge period and the duty cycle pause, both in milliseconds.

TABLE 8.4 (CONTINUED)

TIRIS Bus Protocol Group 2 Commands

---

^d Command $43 (*Set antenna*) selects the active antenna in those interrogators with two antenna ports. This command generates an immediate or broadcast response. It requires a parameter (0 or 1) to select the active antenna.

^e Command $44 (*Get antenna*) returns the currently selected antenna. This command can also generate an immediate or broadcast response. The response is either 0 or 1 indicating the active antenna.

^f Command $45 (*Write EEPROM*) writes one or more bytes into the EEPROM in the interrogator. This command can generate an immediate, broadcast, or queued response. The command requires two parameters: the address to start the writing process and the data to be written.

^g Command $46 (*Read EEPROM*) reads one or more bytes from the EEPROM in the interrogator. This command can also generate an immediate, broadcast, or queued response. The command requires two parameters: the start address in the EEPROM and the number of bytes to be read. The response contains the read data.

^h Command $49 (*Read byte from input*) reads the status of one of the input/output ports in the interrogator, specified by the port identifier. This command can generate an immediate or broadcast response. It requires the identification of the port. The response contains the status of the port.

ⁱ Command $4D (*Write byte to output*) writes a value to the output port of the interrogator specified by the port identifier. The command can generate an immediate or broadcast response. It requires three parameters: the port identifier, the port value, and the logical operation to perform. The output value is the result of the logical operation of the current port setting and the parameter passed to the function:

$00 = write port value direct to the output.

$01 = read status ANDed with port value.

$02 = read status ORed with port value.

$03 = read status exclusive-ORed (XORed) with port value.

$04 = invert current port setting.

^j Command $55 (*Get count of buffered ID records*) returns the count of buffered ID records. This command can only generate an immediate response. The command does not require any parameter. The output is the number of buffered records.

^k Command $56 (*Set read mode*) sets the read mode in the interrogator. This command can generate an immediate, broadcast, or queued response. It requires a single parameter that is the following read mode: 0 = idle mode, and 1 = gate mode. Gate mode is a charge-only function to interrogate transponders continuously without having to execute a single read command each time.

^l Command $57 (*Delete ID records*) deletes a specific number of ID records from the ID memory. It is necessary to specify the start record and the number of records to be deleted. The command generates an immediate or broadcast response.

^m Command $58 (*Get read mode*) returns the current state selected by read mode. This command generates an immediate response. It does not require any parameters. The response contains the currently selected read mode.

ⁿ Command $5A (*Get ID records from the ID buffer*) returns the number of ID records stored in the buffer specified in the command. This command generates an immediate response. The command uses three parameters: the starting record, the number of records, and the length of the record. The response of the command has a number of fields equal to the number of ID records requested plus one. The first field specifies the actual number of records being transmitted, while the rest of the fields contain the ID records. The initial field is required because if the number of records requested exceeds the length of the total records, the command will only transmit the available number of records.

(Continued)

TABLE 8.4 (CONTINUED)

TIRIS Bus Protocol Group 2 Commands

o Command $5B (*Put ID records into the ID buffer*) writes a given number of ID records into the buffer. This command also generates an immediate response only. The command requires the following three parameters: the start record, count of records, and length of the ID records. This is followed then by the ID records to be written in the buffer. This command does not return any data in its response.

p Command $5D (*Read block of data from data memory*) returns a given number of bytes from the data memory in the interrogator. This command generates an immediate response. The command requires two parameters: the start address and the number of bytes to be read. The response contains the contents in the selected memory areas.

q Command $5E (*Write block of data to the data memory*) writes a specified number of data records into the data memory in the interrogator. This command generates an immediate or a broadcast response. It requires as parameters the different data records to be written. The command does not return any data in its response.

r Command $5F (*Reset reader*) resets the interrogator unit. This command generates an immediate or a broadcast response. It does not require any parameters or return any data.

As mentioned in Table 8.5's notes, Table 8.6 shows the relationship between the ASCII character following the *set outputs* command and its effects on the output lines.

Example 8.2:

Describe the effect of sending the character ***1*** after the character ***Y***.

Solution:

The character *Y* triggers the set outputs command. The character *1* specifies the status of the output lines as follows:

Output line 4 → 1
Output line 3 → 0
Output line 2 → 0
Output line 2 → 0

The commands for use with 64-bit read/write transponders are listed in Table 8.7.

The commands for use with the multipage transponder are listed in Table 8.8.

As mentioned in the notes for Table 8.8, Table 8.9 shows that the selection of the receiving antenna is done with input/output lines 6 and 7, while the selection of the transmitting antenna is done with input/output lines 4 and 5.

Example 8.3:

Describe the operation of command sequence LM4201.

Solution:

The first two characters put the interrogator into line-multiplexed mode. The interrogator will read continuously page 1 of a multipage transponder, switching the

TABLE 8.5

ASCII Bus Protocol General Commands

Command Character	Command Name
B	Readout buffer[a]
C	Clear[b]
F	Format[c]
H	Set outputs and get input-output status[d]
J	Input-output Status[e]
V	Version[f]
Y	Set outputs[g]
Z	Set charge period[h]

[a] *Readout buffer* command is triggered by the ASCII character B. With this, the interrogator transmits the character B ($42) and the contents of the buffer to the host computer. This is followed by a carriage return ($0D) and line feed ($0A) characters. If the buffer is empty, the interrogator only sends the character B followed by $0D and $0A.

[b] *Clear* command is triggered by the ASCII character C. This command erases the identification buffer, and the ID counter is reset to zero. After performing this task, the interrogator sends back the ASCII character C ($43) followed by a carriage return ($0D) and the line feed ($0A).

[c] *Format* command is triggered by the ASCII character F. Sending this command to the interrogator causes a change in the output format for page and identification numbers from decimal to hexadecimal. The interrogator transmits back the ASCII character F ($46) followed by a carriage return ($0D) and line feed ($0A) characters once the command has been accepted. The format will continue being hexadecimal until the *escape* command ($1B) is received.

[d] The ASCII character H combines and extends two specific commands: the *I/O status* command and *set outputs* command. After the interrogator receives the ASCII character H, it echoes it back to the host computer, indicating that it is ready to receive further parameters that are needed to specify the operation of the input/output ports. These parameters are the ASCII character I to indicate input ports followed by a number from 0 to 4 that specifies the port, or the ASCII character O to indicate output ports followed by the port specified (0 to 3), the logical operation, and the port value.

[e] *Input/output status* command is triggered by the ASCII character J. This command causes the interrogator to transmit the current status of the four lower I/O lines and the four higher I/O lines. Originally, the interrogators are configured with lines 0 to 3 as inputs and 4 to 7 as outputs, although these can be changed using the appropriate commands. When the interrogator receives the ASCII character J ($4A), it echoes it back to the host computer followed by the ASCII characters that represent the status of the input and output lines. These are followed by a carriage return ($0D) and the line feed ($0A).

[f] *Version* command is triggered by the ASCII character V. This causes the interrogator to transmit the version number of the software, followed by a carriage return ($0D) and the line feed ($0A).

[g] *Set outputs* command is triggered by the ASCII character Y followed by an ASCII character in the range from 0 to F. For the command to operate correctly, one of the I/O line groups must be set to output. The command reads the configuration information and sets the output level of the lines that are considered outputs. In the case that both sets of lines are configured as outputs, the command only acts on the higher lines. If both sets of inputs are configured as inputs, the command does not have any effect. Table 8.6 shows the relationship between the ASCII character following the *set outputs* command and its effects on the output lines.

(Continued)

TABLE 8.5 (CONTINUED)

ASCII Bus Protocol General Commands

ʰ *Set charge period* command is triggered by ASCII character Z. This command sets up the charge time in the range of 15 ms to 255 ms in 1 ms increments. The character Z is followed by two further ASCII characters (in the range from 0 to F) that specify the charge time once the hexadecimal number has been converted into a decimal. The resulting number sets the charge time. For example, sending the characters *64* results in a charge time of 100 ms, and sending the characters *3C* results in a charge time of 60 ms.

TABLE 8.6

Effect of ASCII Character Sent After Set Outputs Command

ASCII character	I/O line 7	I/O line 6	I/O line 5	I/O line 4
0	0	0	0	0
1	0	0	0	1
2	0	0	1	0
3	0	0	1	1
4	0	1	0	0
5	0	1	0	1
6	0	1	1	0
7	0	1	1	1
8	1	0	0	0
9	1	0	0	1
A	1	0	1	0
B	1	0	1	1
C	1	1	0	0
D	1	1	0	1
E	1	1	1	0
F	1	1	1	1

control output lines for a four-channel transmit multiplexer and a two-channel receive multiplexer.

8.2 Commands for Texas Instruments High-Frequency Interrogators

8.2.1 Commands Supported by the TIRIS Protocol for Series 6000 Interrogators

The Series 6000 Reader System, described in Chapter 5, is a family of RFID interrogators operating in the HF band, manufactured by Texas Instruments. The data for the communication from the interrogator to the transponder are known as the *request packet*, while the data used in the communication from the transponder to the interrogator are known as the *response packet*.

TABLE 8.7

ASCII Bus Protocol Commands for Read/Write Transponders

Command Character	Command Name
Esc	Normal mode or escape[a]
G	Gate mode[b]
L	Line function[c]
N	Number[d]
P	Program[e]
R	RAM fill[f]
S	Store[g]
U	Antenna[h]
X	Execute[i]
?	Read memory[j]

[a] *Normal mode* or the *escape* command is triggered by the ASCII character *escape*. The normal function provides a continuous readout of transponders at high speed. When this mode is used, the valid IDs received are compared against the identification already in the buffer. If the ID received is a new one, it overwrites the buffer and then transmits the information to the host computer. The transfer consists of the transponder type character, space, application code, space, identification code, carriage return, and line feed. The ASCII character *escape* ($1B) activates the readout function. The ASCII character *E* ($45), a carriage return ($0D), and the line feed ($0A) are returned by the interrogator.

[b] *Gate mode* command is triggered by the ASCII character G. In this mode, each correct identification number is compared with the identification stored in the identification memory. When a new identification number is received, it is then stored in memory. The interrogator confirms the reception of the command by sending back the ASCII character G ($47), followed by a carriage return ($0D) and the line feed ($0A). Each time a new identification number is detected, the ASCII character G ($47), the transponder type character, space, the actual memory count number, space, the application code, space, the identification code, a carriage return, and the line feed are transmitted via the interface.

[c] *Line function* command is triggered by the ASCII character L. This command switches the interrogator into a special continuous readout mode in which each correct identification causes the direct transmission of the ASCII character *L* ($4C), the transponder type character, space, the application code, space, the identification code, a carriage return, and the line feed without doing any comparison in the buffer. In the case that no identification is received or, when received, it is not valid, the ASCII character *L* together with the carriage return ($0D) and line feed ($0A) characters are transmitted.

[d] *Number* command is triggered by the ASCII character N. This command is used by the host computer to check the number of stored identifications. After receiving it, the interrogator transmits the ASCII character N ($4E), space, the number of identifications in the memory in hexadecimal format, followed by a carriage return ($0D) and line feed ($0A).

[e] *Program* command is triggered by the ASCII character P. This command is used to program a read/write transponder. The host computer first sends the ASCII character P to the interrogator. When the character is echoed, the host computer transmits the 16 hexadecimal characters that specify the identification number to be programmed into the transponder. After the interrogator receives the 16 characters, it performs a block check character (BCC) calculation. Then, the transponder is charged up, and the identification data including the BCC are sent to the transponder.

(Continued)

TABLE 8.7 (CONTINUED)

ASCII Bus Protocol Commands for Read/Write Transponders

Once the programming cycle is complete, the transponder sends the programmed data back to the interrogator to compare it with the original string. After the comparison is done, the interrogator sends a status character followed by a carriage return ($0D) and line feed ($0A) to the host computer. The meanings of the different status characters are as follows:

0 = transponder correctly programmed.

1 = ID received from transponder different from ID transmitted.

2 = interrogator did not receive any ID from the transponder.

It is important to note that the programming cycle is longer than regular reading cycles. For this reason, there may be collisions with other interrogators in the area. Also, it is important to note that the programming distance is lower than the reading distance due to the stronger power requirements during programming.

f *RAM fill* command is triggered by the ASCII character R. This is used by the host computer to test the memory in the interrogator by filling it with predetermined codes and reading it to check for consistency. Two ASCII characters (in the range from 0 to F) are used to specify an 8-bit test pattern. After the memory is filled, the interrogator transmits the ASCII character R ($52), followed by a carriage return ($0D) and the line feed ($0A) to the host computer.

g *Store* command is triggered by the ASCII character S. This command causes the transmission of all stored identification numbers from the interrogator's memory to the host computer together with their memory addresses. Each transmitted line consists of the transponder type, space, the memory location, space, the application code, space, the identification code, a carriage return, and the line feed. After the transmission has finished, the interrogator transmits the ASCII character S, carriage return, and line feed.

h *Antenna* commands are triggered by the ASCII character U. This command allows the user to choose a specific antenna in the interrogator. This is done by transmitting the character 0, 1, or 2 after the character U. The character 1 selects Antenna 1, the character 2 selects Antenna 2, and the character 0 toggles the selected antenna. If the character transmitted after the character U is different from 0, 1, or 2, the whole command is ignored.

i *Execute* command is triggered by the ASCII character X. This causes the interrogator to switch to a single readout mode and triggers a single readout sequence. After the command is executed, the interrogator remains in an idle loop waiting for the next command. After receiving an identification number from a transponder, the interrogator transmits the character X, the transponder type character, space, the four-digit application code, space, the 16-digit identification code, a carriage return, and the line feed to the host computer. If the interrogator does not read an identification number, the interrogator transmits the character X ($58) followed by a carriage return ($0D) and the line feed ($0A). If the interrogator received an identification number but this was invalid, it transmits the character X ($58), the character I ($49), followed by a carriage return ($0D) and line feed ($0A).

j *Read memory* command is triggered by the ASCII character ? After echoing back this character, the interrogator transmits space, the memory location, space, the 16-digit identification code, the transponder type information, a carriage return, and the line feed

TABLE 8.8

ASCII Bus Protocol Commands for Multipage Transponders

Command Character	Command Name
Esc	Normal mode[a] or escape
L	Line function[b]
O	Lock page[c]
P	Program[d]
U	Antenna[e]
X	Execute[f]

Note: Before being able to access multipage transponders, the interrogator must be configured for this operational mode. This allows the interrogator to generate a protocol that includes additional information such as the page in the multipage transponder as well as additional information concerning a particular command. The interrogator enters multipage mode after receiving the character *K* followed by the character *1*. Similarly, the interrogator reverts back to a read/write transponder mode after receiving *K 0*. The interrogator echoes these characters followed by the carriage return and line feed characters.

[a] The *normal* mode is triggered by the ASCII character *escape* followed by a page number. The ASCII character *E*, carriage turn, and line feed are echoed back to the host computer by the interrogator to acknowledge the reception of the command. Normal mode provides a continuous readout of a certain transponder page at high speed. When reading a multipage transponder, it will send back the antenna number, the read status, the transponder type character, space, the page, the pace, the application code, space, the identification code, a carriage return, and the line feed.

[b] *Line function* is triggered by the ASCII character *L*. If this character is followed by a page number, the system continuously reads the specified page. If the character *L* is followed by the character *M*, the interrogator is switched into multiplexed mode. When the transponder is continuously reading the specified page, each correct identification is transmitted directly to the host computer without comparing the received value against a value in the buffer. In addition, before each identification number, the interrogator transmits the ASCII character *L*, the antenna number, the read status information, and the transponder type. If the interrogator does not receive any identification, or the received value is not valid, it transmits the ASCII character *L*, the antenna port used, a carriage return, and the line feed. For the interrogator to operate in multiplexed mode, after the characters *L* and *M*, it must receive the following additional parameters: the number of transmit multiplex antennas, the number of receive multiplex antennas, and the page information to read a multipage transponder. The selection of the receiving antenna is done with input/output lines 6 and 7, while the selection of the transmitting antenna is done with input/output lines 4 and 5, as shown in Table 8.9.

[c] *Lock page* command is triggered with ASCII character *O*. This command enables the user to disable for programming a specific page of a multipage transponder. After the interrogator receives the character *O*, it echoes it back to the host computer, indicating that it is now ready to receive the page number that will be locked. The status command can be used to verify that the page has been successfully locked. When the interrogator has been able to communicate with the transponder, it sends the following information to the host computer: antenna number, status, transponder type, space, page, space, application code, space, identification number, carriage return, and line feed. If the interrogator has not been able to communicate with the transponder effectively, it sends the following information to the host computer: antenna number, carriage return, and line feed.

(Continued)

TABLE 8.8 (CONTINUED)

ASCII Bus Protocol Commands for Multipage Transponders

ᵈ *Program* command allows programming a multipage transponder. To initiate the programming process, the interrogator must receive the ASCII character *P*, two characters containing the page number ($01 to $11), and 16 additional hexadecimal characters containing the ID to be programmed into the specified page. After the interrogator receives the last character from the host computer, it performs a block check character (BCC) calculation and sends the entire string to the transponder. Once the programming has finished, the interrogator receives the identification number and the address from the transponder to be compared with the data sent to the transponder. If the comparison is successful, the interrogator transmits the following information to the host computer: antenna number, programming status, transponder type, space, page, space, application code, space, identification number, carriage turn, and line feed. If the comparison fails, it transmits the antenna number, status information 2, carriage return, and line feed.

ᵉ *Antenna* command is triggered by the ASCII character *U* followed by a character in the range from *0* to *2*. This command allows the user to specify a particular receiving antenna. The character *1* selects Antenna 1, the character *2* selects Antenna 2, and the character *0* toggles the selected antenna. If the character transmitted after the character *U* is different from *0*, *1*, or *2*, the whole command is ignored.

ᶠ *Execute* command is triggered by the ASCII character *X* followed by a page number ($01 to $11) or the character *M*. When the characters $01 to $11 follow the character *X*, the interrogator operates in a single readout mode, reading the page specified by the $01 to $11 characters. When the character *M* follows the character *X*, the interrogator enters into a multiplexed mode operating similarly to its operation mode for a read/write transponder.

The commands supported by this protocol are divided into the following functional categories:

- Tag-it HF commands
- Miscellaneous commands
- ISO/IEC 15693-3 commands

Table 8.10 shows the different Tag-it HF commands. These commands are specific for working with the Tag-it family of HF transponders. These commands can be applied to addressed and nonaddressed blocks.

The commands in the *miscellaneous* category are shown in Table 8.11.

The packet for the request (interrogator to transponder) consists of a header, the packet length, the node address, command flags, the interrogator command ($60), the ISO/ECI 15963-3 command, and data bytes 0 to *n* and checksum, as shown in Table 8.12. Table 8.13 describes its contents.

For the commands that follow the ISO/IEC 15963-3, the command field always contains $60. The actual command to be executed by the interrogator is contained in the data field. The response packet has a similar structure; it consists of a header, the packet length, the node address, command flags, the interrogator command ($60), the ISO/IEC 15963-3 command and data bytes, and a checksum, as shown in Table 8.14. Table 8.15 describes the contents of the response packet.

TABLE 8.9

Selection of Antennas in Multiplexed Mode

RX Antenna Controlled with I/O Lines 6 and 7		
	I/O 6	I/O 7
RX antenna 1	0	0
RX antenna 2	1	0
RX antenna 3	0	1
RX antenna 4	1	1

TX Antenna Controlled with I/O Lines 6 and 7		
	I/O 4	I/O 5
TX antenna 1	0	0
TX antenna 1	1	0
TX antenna 1	0	1
TX antenna 1	1	1

For commands that follow the ISO/IEC 15963-3 protocol, the command field is always equal to $60. Table 8.16 shows the commands that comply with the ISO/IEC 15963 Part 3 (15963-3) transmission protocol.

For Table 8.16, Commands $01 and $02 are considered *mandatory commands*. Command $01 is the *inventory command*. Figure 8.1 shows the format of this command. It can be seen how the request command packet follows the format outlined in Figure 8.1, while the specifics for this command are included into the ISO command data field. The packet format as a response to the inventory command is shown in Figure 8.2.

The field *valid data flags* contains 16 bits that specify whether or not each one of the possible 16 time slots received valid data. Bits 0–7 in the LSB byte correspond to time slots 1 to 8, while bits 0 to 7 in the MSB byte correspond to time slots 9 to 16. The reception of valid data is indicated when the appropriate bit is set to 1. The field *collision flags* contains 16 bits that specify whether or not a collision occurred in the possible 16 time slots. The structure of these bits follows the structure described for the valid data flags. A collision in a specific time slot is indicated by the appropriate bit set to 1.

Command $02 is the *ISO stay quiet request* command. After a transponder following the ISO 15963-3 protocol receives this command, it enters in quiet state and will not initiate a response. The request format follows the same structure as the other ISO commands, with the ISO command data shown in Figure 8.3. It is necessary to note that after receiving this command, the transponder does not issue a response.

The rest of the ISO commands are considered optional commands. The structure of their diverse command data fields follows the standard structure with each specific command code.

TABLE 8.10

Tag-it HF Commands for TIRIS Protocol Series 6000

Command Code	Command Name
$02	Read block[a]
$03	Write block[b]
$04	Lock block[c]
$05	Read transponder details[d]
$0F	Special read block[e]

[a] Command code $02 (*read block*) reads a single block of data from an HF Tag-it transponder. When the address flag is set, the address is part of the data section with its least significant bit (LSB) first. This is followed by a byte that contains the number of the block to be read. When the address flag is clear, the address is not sent.

[b] Command code $03 (*write block*) writes the specified data into a block in an HF Tag-it transponder. When the address flag is set, the address is part of the data section with its LSB first. This is followed by a byte that contains the number of the block to be read. The data to be written follow the block number. When the address flag is clear, the address is not sent and only the block and data are being sent. The response follows a similar structure in which the data have been changed to $00 in order to indicate a successful operation.

[c] Command code $04 (*lock block*) locks a single block of data in an HF Tag-it transponder. When the address flag is set, the address forms the first part of the data section. This is followed by a single byte containing the number of the block to lock. The response contains $00 in the data section, indicating a successful lock operation.

[d] Command code $05 (*read transponder details*) reads the details of an HF Tag-it transponder. The data section in the response contains the transponder address, the manufacturer code, the version number of the transponder, the number of blocks, and the number of bytes per block.

[e] Command code $0F (*special read block*) reads blocks of data from an HF Tag-it transponder. The data section contains a single byte that specifies the number of blocks to be read. Each bit in this byte represents a block of data. Each bit that is set to 1 marks that block number to be read.

8.2.2 Commands Supported by HF Tag-It Transponders

The following describes the commands supported by the HF Tag-it transponders manufactured by Texas Instruments. While some of these commands may be used in a nonaddressed mode, it is necessary to be extremely careful when using this method, as it may produce undesired effects in nearby transponders. For example, a *lock block* command executed in a nonaddressed mode will effectively lock all the transponders within its write range. For this reason, it is recommended to specify the specific address of the transponders.

Command $01 (*read block*) reads the contents of a single block of data from a transponder. The request message has two fields: 4 bytes for the address field and 1 byte for the number of the block to be read. The data in the response message have the fields shown in Table 8.17.

TABLE 8.11

Miscellaneous Commands for TIRIS Protocol Series 6000

Command Code	Command Name
$D0	Initiate FLASH Loader[a]
$D8	Send Data to FLASH[b]
$F0	Reader Version[c]
$F1	Reader Inputs[d]
$F2	Write Reader Outputs[e]
$F4	RF Carrier[f]
$FF	Baud Rate[g]

[a] Command code $D0 (*initiate Flash loader*) is used to initialize and transfer control to the Flash loader software. The response of the system contains $00 in the data place if the command has been successful.

[b] Command code $D8 (*send data Flash*) is used to load up to 132 bytes of data into the Flash memory. The data bytes are followed by 2 checksum bytes. If the command is successful, the response contains $00 in the data place.

[c] Command $F0 (*reader version*) requests the version of the interrogator. The response is similar to the request packet, with the data section containing the 2-byte version LSB first followed by a single byte representing the type of interrogator.

[d] Command code $F1 (*reader inputs*) reads the state of the different interrogator's inputs. The data section in the response contains a byte representing the state of the inputs, with bit 0 representing input 1 and bit 1 representing input 2.

[e] Command $F2 (*write reader outputs*) writes the state of the interrogator's outputs. The meaning of the different bits for the 1-byte data being sent is as follows:

Bit 0   1 = Output 1 pulled to ground
Bit 1   1 = Output 2 pulled to ground
Bit 2   Reserved
Bit 3   Reserved
Bit 4   1 = Bit 0 enabled
Bit 5   1 = Bit 1 enabled
Bit 6   Reserved
Bit 7   Reserved

[f] Command $F4 (*RF carrier*) switches the radiofrequency carrier on and off. The carrier is turned on by loading the $FF into the data section and turned off by loading $00 into the data section.

[g] Command $FF (*baud rate*) sets up the baud rate for the interrogator. The data section consists of 1 byte linked to the baud rate as follows:

09 = 57,600 baud
08 = 38,400 baud
07 = 19,200 baud
06 = 9600 baud

In addition to supporting the proprietary Tag-it protocol, the Series 6000 interrogators also comply with the ISO/IEC 15963 protocol. Under this protocol, the communication from the interrogator to the transponder uses amplitude-shift keying (ASK) at a modulation rate of either 100% or between 10% and 30%. The response from the transponder to the reader uses frequency-shift keying (FSK). These parameters are set by the configuration byte that is ISO command data byte 0.

TABLE 8.12

Structure of the Request Packet for TIRIS Protocol Series 6000

Header	Packet Length		Address Node		Command Flag	Command	Command Data		Checksum	
$01	LSB	MSB	LSB	MSB	Flags	$60	$XX Data		Byte1	Byte2
1 byte	2 bytes		2 bytes		1 byte	1 byte	Byte0 Bytes 1–n		2 bytes	
							1 byte N bytes			

TABLE 8.13

Format of Request Packet for TIRIS Protocol Series 6000

Field	Length	Description
Header	1 byte	Start of packet $01
Packet Length	2 bytes	Length of packet. Includes checksum
Node Address	2 bytes	Interrogator's address
Command Flags	1 byte	Specifies execution of command
Command	1 byte	Command for interrogator to execute
Data	0 to n bytes	Data required to execute command
Checksum	2 bytes	Byte 1 is XOR of all elements from header to last byte. Byte 2 is calculated XORing Byte 1 with $FF.

TABLE 8.14

Structure of the Response Packet for TIRIS Protocol Series 6000

Header	Packet Length		Address Node		Command Flag	Command	Command Data	Checksum	
$01	LSB	MSB	LSB	MSB	Flags	$60	Data	Byte1	Byte2
1 byte	2 bytes		2 bytes		1 byte	1 byte	m bytes	2 bytes	

TABLE 8.15

Format of Response Packet for TIRIS Protocol Series 6000

Field	Length	Description
Header	1 byte	Start of packet $01
Packet Length	2 bytes	Length of packet. Includes checksum
Node Address	2 bytes	Interrogator's address
Command Flags	1 byte	Response of interrogator to request. Bit 4 defines the error status. If bit 4 = 1, error has occurred.
Command	1 byte	Command that the interrogator executed
Data	0 to m bytes	Data returned by interrogator in response to command
Checksum	2 bytes	Byte 1 is XOR of all elements from header to last byte. Byte 2 is calculated XORing Byte 1 with $FF.

TABLE 8.16

ISO/IEC 15963-3 Protocol

Command Code	Command Name
$01	Inventory
$02	Stay Quiet
$20	Read Single Block
$21	Write Single Block
$22	Lock Block
$23	Read Multiple Blocks
$27	Write AFI
$28	Lock AFI
$29	Write DSFID
$2A	Lock DSFID
$2C	Get Multiple Block Security Status

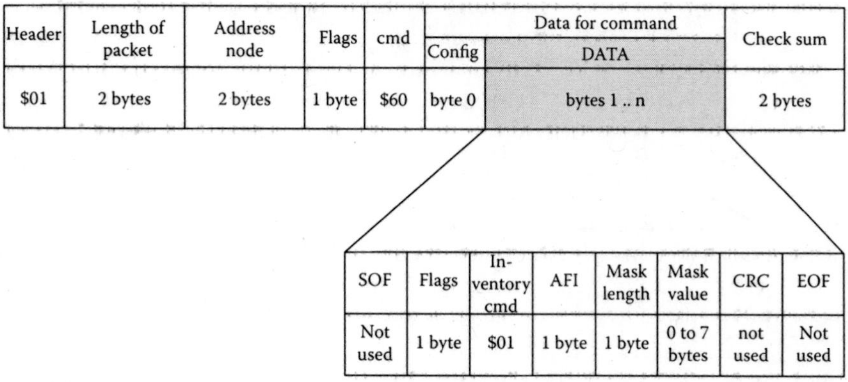

Header	Length of packet	Address node	Flags	cmd	Data for command		Check sum
					Config	DATA	
$01	2 bytes	2 bytes	1 byte	$60	byte 0	bytes 1 .. n	2 bytes

SOF	Flags	In-ventory cmd	AFI	Mask length	Mask value	CRC	EOF
Not used	1 byte	$01	1 byte	1 byte	0 to 7 bytes	not used	Not used

FIGURE 8.1
Structure of the request packet when sending the *inventory* command.

Command $02 (*read multiblock*) reads the contents of more than one block of data from a transponder. The request message contains three fields: the 4-byte address for the transponder, the number of the first block to be read, and the number of blocks to be read. The data in the response message follow the format shown in Table 8.18.

Command $03 (*tag version*) reads the information from a transponder that was programmed at the factory and cannot be altered. The request message has a single field that contains the 4-byte address for the transponder being interrogated. The data in the response message follow the format shown in Table 8.19.

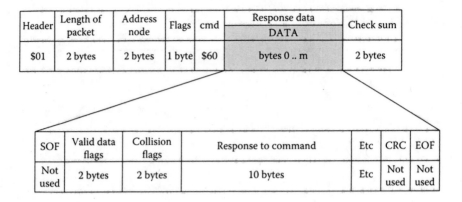

FIGURE 8.2
Structure of the response packet for the *inventory* command.

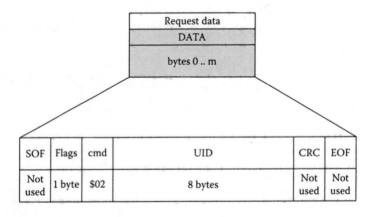

FIGURE 8.3
Request data for the *stay quiet* command.

TABLE 8.17

Format of Data Response for Read Block Command

Field Name	Number of Bytes	Description	Comments
Address	4	Transponder address	
Block Number	1	Number of block to read	$00 to $07
Lock Status	1	Status of block lock	$00 = unlocked $01 = locked by user $02 = locked by factory
Block Data	4	Data	

TABLE 8.18

Format of Data Response for Read Multiblock Command

Field Name	Number of Bytes	Description	Comments
Address	4	Transponder address	
Number of blocks read	1	Specified the number of times that the next three fields are repeated	
Block number	1	Number of block	$00 to $07
Lock status	1	Status of block lock	$00 = unlocked $01 = locked by user $02 = locked by factory
Block data	4	Data	
Error number	1	Field will be present if not all requested blocks were read	Specific error codes: $11 = At least one block not available $15 = At least one block not read

TABLE 8.19

Format of Data Response for Tag Version Command

Field Name	Number of Bytes	Description	Comments
Address	4	Transponder address	
Manufacturer Code	1	Identifies manufacturer of transponder	Only 7 LSBs
Chip Version	2	Chip version used	Only 9 LSBs
Block Size	1	Size of each block for this transponder	Counts on bytes starting from zero. Only 5 LSBs
Number of blocks	1	Number of blocks in transponders	Counts starting from zero

More general commands:

- Command $04 (*stop continuous*) causes the interrogator to stop operating in continuous mode. This command has no effect if the interrogator is not operating in continuous mode. There are no data associated with this command.

- Command $05 (*write block*) writes a single block of data to the transponder. The request message contains three fields: the 4-byte address for the transponder, the number of the block to be written, and the 4-byte data to write. The data in the response message contain a single field that is the transponder address.

- Command $06 (*write multiblock*) writes data to more than one block in the transponder. The request message contains four fields: the 4-byte address for the transponder, the number of the first block, the number of blocks, and the 4-byte data to write. The data in the response message contain two fields: the 4-byte transponder address and an optional field with the error number. This field is not present if the command was executed without errors. Some specific error codes are as follows:

$11 = At least one block not available

$13 = At least one block already locked

$17 = At least one block not successfully written

- Command $07 (*write and lock block*) writes a single block of data to a transponder and locks it. The request message contains three fields: the 4-byte address for the transponder, the number of the block to be written, and the 4-byte data to write. The data in the response message contain only one field that is the transponder address.

- Command $08 (*lock block*) locks a single block of data in a transponder. The request message contains two fields: the 4-byte address of the transponder and the block number. The data in the response message contain a single field that is the transponder address.

- Command $09 (*lock multiblock*) locks more than one block of data in a transponder. The request message contains three fields: the 4-byte address of the transponder, the first block number, and the number of blocks to lock. The data in the response message contain two fields: the 4-byte transponder address and an optional field with the error number. This field is not present if the command was executed without errors. Some specific error codes are as follows:

$11 = At least one block not available

$13 = At least one block already locked

$19 = At least one block not successfully locked

- Command $10 (*reset reader*) causes the interrogator to perform a software reset. Prior to executing the reset, the interrogator completes the command that is currently executing. After the reset, any temporary settings or data stored in volatile memory are lost. There are no data associated with this command.

- Command $11 (*reader version*) makes the interrogator send the version of the firmware that is running. There are no data associated with this request. The data in the response message contain two fields: a 3-byte field for the firmware version and a 1-byte field for the firmware type. This can be either the standard version (bit 0 = 0) or engineering version (bit 0 = 1).

- Command $12 (*reader diagnostic*) makes the interrogator perform self-diagnosis operations. The request message contains the 1-byte operation to be performed following this code:

$01 = Self-test

$02 = Reserved

$03 = Transmitter ON

$04 = Transmitter OFF

$05 = Detailed error messages

$06 = Normal error messages

The response contains a single 1-byte field that describes the status of the interrogator after the execution of the operation:

$01 = Self-test failed

$FF – Self-test passed

$02 = Reserved

$03 = Transmitter turned ON

$04 = Transmitter turned OFF

$05 = Detailed error messages ON

$06 = Normal error messages ON

- Command $13 (*read reader setup*) makes the interrogator send the current setup and configuration data from its on-board Flash memory. There are no data associated with this request. The data in the response message are a single field of up to 12 bytes containing the data stored in its on-board memory.
- Command $16 (*start Flash loader*) is used to reprogram the Flash memory, for firmware upgrades, or to set up data alteration. After the Flash loader has started, the interrogator ceases communicating using the host protocol and instead uses the one from the loader. The Flash loader is deactivated by a soft or a hard reset. There are no data associated with this request or a response.
- Command $3D (*factory lock block*) locks a single block of data on a transponder. This is a factory command not available to general users.
- Command $3E (*write SID code*) stores the session ID (SID) code into the transponder. This SID code becomes the new address for the

transponder. The request message contains a single field with the 4-byte address of the transponder. The data in the response message also contain the address of the transponder.

- Command $3F (*factory programming off*) clears the factory programming bit, in turn disabling all factory programming functions. This is also a factory command not available to general users.

- Command $FD (*read multiblock SID*) reads the contents of more than one block of data from all the transponders within reading range. This command cannot be executed in addressed mode. The maximum number of transponders that is possible to read in a single pass is 100. The request message contains three fields: the first block number to be read, the number of blocks, and a field indicating the options for the response. The data in the response message follow the format shown in Table 8.20.

- Command $FE (*read block SID*) reads the contents of a single block of data from all the transponders within reading range. This command cannot be executed in addressed mode. The maximum number of

TABLE 8.20

Format of Data Response for Read Multiblock SID Command

Field Name	Number of Bytes	Description	Comments
Response Options	1	Specifies how data will be returned	Bit 0: 0 = do not send addresses 1 = show addresses Bit 1: 0 = no version data 1 = version data
Address	4	Transponder address (not always present)	
Manufacturer code	1	Identifies manufacturer of transponder	Only 7 LSBs
Chip version	2	Chip version used	Only 9 LSBs
Block size	1	Size of each block for this transponder	Counts on bytes starting from zero. Only 5 LSBs
Number of blocks	1	Number of blocks in transponders	Counts starting from zero
Number of blocks read	1	Determines how many times the following 3 fields will be repeated	
Block number	1	Number of blocks	$00 to $07
Lock status	1	Locked status of block	$00 = unlocked $01 = locked by user $02 = locked by factory
Block data	4	Data	

TABLE 8.21

Format of Data Response for Read Block SID Command

Field Name	Number of Bytes	Description	Comments
Response Options	1	Specifies how data will be returned	Bit 0: 0 = do not send addresses 1 = show addresses Bit 1: 0 = no version 1 = data Bit 2: 0 = response with block data 1 = response with address only
Number of transponders found	1	Number of times the following 3 fields are repeated	
Address	4	Transponder address (not always present)	
Manufacturer code	1	Identifies manufacturer of transponder	Only 7 LSBs
Chip version	2	Chip version used	Only 9 LSBs
Block size	1	Size of each block for this transponder	Counts on bytes starting from zero. Only 5 LSBs
Number of blocks	1	Number of blocks in transponders	Counts starting from zero
Block number	1	Not always present	
Lock status	1	Not always present	
Block data	4	Not always present	

transponders that is possible to read in a single pass is 100, although each single response message carries information for up to 50 transponders. The request message contains two fields, the block number to be read and an options field that determines how the data will be returned. The data in the response message follow the format shown in Table 8.21.

8.3 Command Supported by Texas Instruments Series 4000 Multifunction Interrogators

The Series 4000 transponders from Texas Instruments are able to operate in the HF and LF ranges. For this reason they need to communicate with different types of transponders following different protocols. This section explores the different commands supported by the different transponders that can communicate with the Series 4000 Interrogators.

8.3.1 Commands Supported by Tag-It Transponders

Many of the responses produced by these commands contain a field called *response flag*. The response flag bits, shown in Table 8.22, are the least significant bits in this field. The host computer must take them into account when decoding the response given to the interrogator because their contents vary depending on the settings being used.

- Command $41 (*find token request*) is used by the host computer to check if a transponder is present. If multiple transponders are present, the command returns the first found. The request packet contains three fields: entity ID (set to $05), the command code ($41), and the loop count field that specifies the number of attempts to find a transponder. The first four fields of the response packet are the entity ID (Set to $05), the issued command ($41), the status showing the error codes, and the library entity ID set to $05. The following 16 commands contain the ID of the tokens being detected.

- Command $45 (*pass-through request*) allows the host to have a direct communication with the radiofrequency module in the interrogator. The request packet contains four fields: entity ID (set to $05), the command code ($45), the number of bits in the next field, and the data field containing the communication to the specific module. The response packet has five fields: the entity ID (Set to $05), the issued command ($45), the status showing the error codes, the number of

TABLE 8.22

Response Flag Bits

Bit Number	Flag

0	Error Flag:
	    0 = No error
	    1 = Error

1	Reserved. Set to 0 by protocol

2	Address Flag:
	    0 = Non-addressed
	    1 = Addressed

3	Format type. Set to 0 by protocol
4	Unused. Set to 0
5	Unused. Set to 0
6	Unused. Set to 0
7	Unused. Set to 0

bits for the next field, and the data field that contains the number of bits in the reply if one was issued.

- Command $48 (*transmitter ON request*) turns on the transmitter in the interrogator for a specific entity that is specified in the request packet. The request packet contains two fields: the entity ID ($05) and the issued command ($48). The response packet contains three fields: the entity ID (set to $05), the issued command ($48), and the status showing the error codes.

- Command $49 (*transmitter OFF request*) turns off the transmitter for a specific identity that is specified in the request packet. The request packet contains two fields: the entity ID ($05) and the issued command ($49). The response packet contains three fields: the entity ID (Set to $05), the issued command ($49), and the status showing the error codes.

- Command $61 (*get block request*) gets data from one memory block of the responding transponder. The request packet contains four fields: the entity ID ($05), the issued command ($61), and the block number and ID of the transponder. In addition to the data read from the transponder, the response packet also contains a block security status field that indicates the write status of the block specified in the request. The structure of the response packet is shown in Table 8.23.

- Command $62 (*get IC version request*) returns information regarding the integrated circuit in the transponder such as version, manufacturer, memory size, number of blocks, and so on. The request packet contains three fields: the entity ID ($05), the issued command ($62), and the ID of the transponder. The response packet contains four fields: the entity ID (Set to $05), the issued command ($49), the status field with the standard error codes, and the response data. The

TABLE 8.23

Structure of Response Packet for Get Block Command

Field	Bytes	Value (Range)	Comments
Cmd 1	1	$05	Entity ID
Cmd 2	1	$61	Get block command
Status	1	$00 to $FF	Error codes
Command code	1	$01	Get block
Response Flag	1	$00 to $04	Response flags
SID	0 to 4	$00 to $FF	ID of transponder
Block Number	1	$00 to $FF	Block number
Lock Status	1	$00 to $03	Write status
Block Data	n	$00 to $FF	Data

response data comprise an 11-byte field in which the contents of the bytes are as follows:

Byte 1: Command code

Byte 2: Response flags

Bytes 3 to 6: ID of transponder

Byte 7: Manufacturer code

Bytes 8 and 9: Chip version

Byte 10: Block bytes minus one

Byte 11: Number of blocks minus one

If the interrogator does not receive a response from the transponder, the last field is substituted by a *no data* field that is zero bytes long.

- Command $63 (*put block request*) writes data to a memory block. To use this command, it is necessary to know the size of the memory block that can be obtained through the *Get IC version* command. Table 8.24 shows the structure of the request packet, and Table 8.25 shows the structure of the response packet.

TABLE 8.24

Structure of Request Packet for Put Block Request Command

Field	Bytes	Value (Range)	Comments
Cmd 1	1	$05	Entity ID
Cmd 2	1	$63	Put block command
Blk Num	1	$00 to $FF	Specifies block to write
BlkBits	1	$00 to $FF	Number of bits next field
BlkData	1 to 32	$00 to $FF	Data to be written
SID	0 to 4	$00 to $FF	ID of transponder

TABLE 8.25

Structure of Response Packet for Put Block Request Command

Field	Bytes	Value (Range)	Comments
Cmd 1	1	$05	Entity ID
Cmd 2	1	$63	Put block command
Status	1	$00 to $FF	Standard Error Codes
CmdCode	0 to 1	$05	Command code
RespFlags	0 to 1	$00 to $05	Response Flags
SID	0 to 4	$00 to $FF	ID of transponder
ErrorResp	0 to 1	$00 to $FF	Error code

- Command $64 (*put block lock request*) writes data to one memory block and locks it to further write operations. The host must also know the size of the memory block. Table 8.26 shows the structure of the request packet, and Table 8.27 shows the structure of the response packet.
- Command $65 (*lock block request*) protects one memory block against writing. If the command is used in a nonaddressed manner without the SID field being present, the command acts on all the transponders in its write zone. The structure of the request packet contains four fields: the entity ID ($05), the issued command ($65), the block number, and the ID of the transponder. The structure of the response packet is shown in Table 8.28.
- Command $66 (*SID poll request*) is used to acquire the IDs of transponders in the interrogator's read area. The possibility of collision among all these transmissions can be reduced by issuing further commands to the transponders to respond in a predetermined slot based on a portion of their ID numbers. Table 8.29 shows the structure of the request packet, and Table 8.30 shows the structure of the response packet.
- Command $67 (*slot marker/end-of-frame request*) is used together with other commands as part of an exchange sequence. The purpose of this command is to send the end-of-frame marker as in the SID

TABLE 8.26

Structure of Request Packet for Put Block Lock Request Command

Field	Bytes	Value (Range)	Comments
Cmd 1	1	$05	Entity ID
Cmd 2	1	$64	Put block lock
Blk Num	1	$00 to $FF	Specifies block to write
BlkBits	1	$00 to $FF	Number of bits next field
BlkData	1 to 32	$00 to $FF	Data to be written
SID	0 to 4	$00 to $FF	ID of transponder

TABLE 8.27

Structure of Response Packet for Put Block Lock Request Command

Field	Bytes	Value (Range)	Comments
Cmd 1	1	$05	Entity ID
Cmd 2	1	$64	Put block lock
Status	1	$00 to $FF	Standard Error Codes
CmdCode	0 to 1	$07	Command code
RespFlags	0 to 1	$00 to $05	Response Flags
SID	0 to 4	$00 to $FF	ID of transponder
ErrorResp	0 to 1	$00 to $FF	Error code

TABLE 8.28

Structure of Response Packet for Lock Block Request Command

Field	Bytes	Value (Range)	Comments
Cmd 1	1	$05	Entity ID
Cmd 2	1	$65	Lock block
Status	1	$00 to $FF	Standard Error Codes
CmdCode	0 to 1	$08	Command code
RespFlags	0 to 1	$00 to $05	Response Flags
SID	0 to 4	$00 to $FF	ID of transponder
ErrorResp	0 to 1	$00 to $FF	Error code

TABLE 8.29

Structure of Request Packet for SID Poll Request Command

Field	Bytes	Value (Range)	Comments
Cmd 1	1	$05	Entity ID
Cmd 2	1	$66	SID Poll Command
ReqVersion	1	$00 to $FF	0 = request only SID in reply
MskLen	1	$00 to $20	Number of bits in next field
MskVal	0 to 4	$00 to $FF	Anti collision maks

TABLE 8.30

Structure of Response Packet for SID Poll Request Command

Field	Bytes	Value (Range)	Comments
Cmd 1	1	$05	Entity ID
Cmd 2	1	$66	SID Poll Command
Status	1	$00 to $FF	Error codes
Command code	1	$0A	Get Phase-Locked Loop (PLL)
Response Flag	1	$00 to $0F	Response flags
SID	4	$00 to $FF	ID of transponder
VersionDat	0 to 5	$00 to $7F	Byte 1: Manufacturer code
		$00 to $FF	Byte 2–3: Chip Version
		$00 to $1F	Byte 4: Block bytes -1
		$00 to $FF	Byte 5: Number of blocks -1

sequence; this marker is understood as a "slot marker" for anticollision. The request packet contains three fields: the entity ID ($05), the issued command ($67), and a command for formatting the reply. The structure of the response packet is shown in Table 8.31.

- Command $68 (*quiet request*) is used to silence a transponder in order to prevent it from responding to any nonaddressed commands

TABLE 8.31

Structure of Response Packet for Slot Marker and End-of-Frame (EOF) Request

Field	Bytes	Value (Range)	Comments
Cmd 1	1	$05	Entity ID
Cmd 2	1	$67	SID Poll Command
Status	1	$00 to $FF	Error codes
Command code	1	$0A	Get Pll
Response Flag	1	$00 to $0F	Response flags
SID	4	$00 to $FF	ID of transponder
VersionDat	0 to 5	$00 to $7F	Byte 1: Manufacturer code
		$00 to $FF	Byte 2–3: Chip Version
		$00 to $1F	Byte 4: Block bytes -1
		$00 to $FF	Byte 5: Number of blocks -1

within its range. The command does not prevent, however, responses to addressed commands. The request packet contains three fields: the entity ID ($05), the issued command ($68), and the SID field. The response packet contains three fields: the entity ID ($04), the command issued (*quiet*), and the status field containing the standard error codes.

8.3.2 Commands Supported by the ISO 14443 Protocol

The ISO 14443 protocol is used by RFID systems operating in the HF range. The protocol is divided into different subparts depending on the anticollision protocols that they use (ISO 14443-3 Type-A and ISO 14443-3 Type-B) and ISO 14443-4 Type-B.

Table 8.32 lists the command supported by the ISO 1444-3 Type-A protocol. While some commands have the same code as those supported by the Tag-it protocol, it is necessary to state that they form parts of separate command sets.

Table 8.33 lists the commands supported by the ISO 14443-3 Type-B protocol.

8.4 Commands for the Texas Instruments UHF Gen 2 Protocol

Table 8.34 lists the different commands supported by the Texas Instruments UHF Gen 2 transponders.

As soon as the integrated circuit in the transponder is immersed in the radiofrequency field generated by the interrogator, it enters the *ready* state and is able to accept the *select* command. The *select* command is sent to all

TABLE 8.32

Commands for ISO 14443-3 Type A

Command code	Command
$41[a]	Find Token Request
$45	Pass-through Request
$48	Transmitter ON Request
$49	Transmitter OFF Request
$61/$62[b]	REQA/WUPA Request
$63[c]	HLTA Request
$64[d]	Anticollision Select Request
$65[e]	RATS (Request Answer to Select) Request
$66[f]	PPS (Protocol and Parameter Selection Request

[a] Commands $41 (*find token request*), $45 (*pass-through request*), $48 (*transmitter ON request*), and $49 (*transmitter OFF request*) operate in the same form as described for Tag-it transponders.

[b] Commands $61 and $62 (*REQA/WUPA request*) are used by the host computer to request to send a REQA/WUPA data packet over the radiofrequency interface. Command $61 requests the REQA packet, while Command $62 requests the WUPA packet. If the command produces data, the response packet contains four fields: the first field is the entity ID for the ISO 14443 ($02), the second field is the requested command ($61 or $62), the third field contains the standard error codes, and the last field contains bytes 1 and 2 for the ATQA (Answer to Request Type A) data from the token.

[c] Command $63 (*HLTA request*) is used by the host computer to request to send an HLTA data packet. The response contains three fields: the first field is the entity ID for the ISO 14443 ($02), the second field is the requested command ($63), and the third field contains the standard error codes.

[d] Command $64 (*anticollision select request*) is used by the host computer to request the transponder to send either an anticollision or a select packet. The response packet contains three common fields: the entity ID for the ISO 14443 ($02), the requested command ($64), and the standard error codes. If the host requested an anticollision packet, the fourth field varies from 2 to 7 bytes and contains the anticollision packet. If the host requested the select data, the fourth field varies from 2 to 5 bytes containing the SAK (Select Acknowledge) data as well as the CRC over the SAK data.

[e] Command $65 (*RATS [request answer to select] request*) is used by the host computer to request the transponder to send RATS data. Before this command can be used, however, the transponder must have been previously activated by a REQA/WUPA, anticollision, or select commands. The response contains the following four fields: the first field is the entity ID for the ISO 14443 ($02), the second field is the requested command ($65), the third field contains the standard error codes, and the last field contains the CID data assigned to the transponder.

[f] Command $66 (*PPS [protocol and parameter selection] request* is used by the host to request a PPS packet. Once again, before this command can be executed, the transponder must have been previously activated. The response contains the following four fields: the first field is the entity ID for the ISO 14443 ($02), the second field is the requested command ($66), the third field contains the standard error codes, and the last field contains the PPS data, including a CRC over the data sent.

TABLE 8.33

Commands for ISO 14443-3 Type-B Protocol

Command code	Command
$41[a]	Find Token Request
$45	Pass-through Request
$48	Transmitter ON Request
$49	Transmitter OFF Request
$61/$62	REQA/WUPA Request
$63[b]	Slot Marker Request
$64[c]	ATTRIB Request
$65[d]	HLTB Request (Halt type B)

[a] The first five commands are similar to those described for the Type-A protocol, and therefore they will not be repeated here. A difference, however, is that the first field in the request and response packets, the entity ID field, is equal to $03 instead of $02 used for Type-A transponders.

[b] Command $63 (*slot marker request*) is used by the host to request the transponder to send a slot marker packet. The number of available slot markers is between 0 and 15. The response packet contains the following four fields: the first field is the entity ID for the ISO 14443 ($03), the second field is the requested command ($63), the third field contains the standard error codes, and the fourth field is a 14-byte field containing the information requested including a CRC.

[c] Command $64 (*ATTRIB [attribute] request*) is used by the host to request to the transponder to send an ATTRIB packet. The response packet contains the following four fields: the first field is the entity ID for the ISO 14443 ($03), the second field is the requested command ($64), the third field the standard error codes, and the fourth field contains the requested information including a CRC over its data.

[d] Command $65 (*HLTB request*) is used to request a HLTB packet. The response packet contains the following four fields: the first field is the entity ID for the ISO 14443 ($03), the second field is the requested command ($65), the third field contains the standard error codes, and the fourth field contains the requested information including a CRC over its data.

the transponders in the interrogator's reading region to inform them that the inventory process will be coming. The *select* command does not require a response from the transponder.

Each one of the inventory rounds starts with a *query* command. This command passes a value from which the transponder generates a slot counter number. If the transponder generates a slot counter equal to zero, it is allowed to reply to the interrogator by sending a 16-bit random number and at the same time entering its *reply* state. The other transponders enter in a waiting mode. After the interrogator has successfully received the response from the transponder, it generates an *acknowledged* (ACK) command and sends it to the transponder along with the same 16-bit number that it first received from the transponder. This allows the transponder to send back its electronic product code (EPC) number and change the state to *acknowledged*. If it is necessary to perform additional operations in the transponder, the interrogator generates a *Req_RN* command, after which

TABLE 8.34

Commands Supported by Texas Instruments UHF Gen 2 Protocol

Command	Binary Code	Operational Group
Select	1011	Select
ACK	01	Inventory
NAK	11000000	Inventory
Query	1000	Inventory
Query_Adjust	1001	Inventory
Query_Rep	00	Inventory
Req_RN	11000001	Inventory
Read	11000010	Access
Write	11000011	Access
Kill	11000100	Access
Lock	11000101	Access
Access	11000110	Access
Block Write	11000111	Access
Block Erase	11001000	Access

the transponder generates and sends a new 16-bit random number called a *handle* and changes its state to *open*. At this point, the transponder is able to accept operations such as *reading, writing, lock,* and *kill*. The exchange terminates when the interrogator sends a *Query_Adjust* command, after which the transponder reverts to ready mode. The *Query_Adjust* command causes to lessen the value of the slot counters in the other transponders in the vicinity. The transponder whose slot counter reaches zero is allowed to respond. This process is repeated until all the transponders in the reading area have been found. If two different transponders were to respond at the same time and the interrogator was unable to identify one of them, they will time out and their slot counters will regenerate. These different commands are as follows:

- The *Select* command is the first message that is sent to the transponder. The parameters that are sent with this command are as follows:
  - *Target*: Instructs the transponder to select one of four sessions and either the SL flag or the Inventory flag associated with the session.
  - *Action*: How to set the flag being selected.
  - *MemBank*: Selects a memory bank.
  - *Pointer*: Where to look in the memory bank.
  - *Length*: How many bits (0 to 255) from the memory bank.
  - *Mask*: Data string that will be compared with the data selected in the memory bank.

- *Truncate*: If a query command specifies Sel = 10 or Sel = 11, the response is truncated to only the EPC data instead of the complete 96 bits.
- *CRC-16*: 16-bit CRC calculated over the command string.

Table 8.35 shows an example of use for the select command.

- The *ACK* command is used by the interrogator to acknowledge the reception of the 16-bit random number sent by the transponder. This command does not need any parameters. When the transponder receives the ACK command, it responds with its EPC data.
- The *not acknowledged* (NAK) command is used to return all the transponders back to the *arbitrary* state. The command does not require any parameter. The command does not generate any responses from the transponders.
- The *query* command initiates the inventory process in order to identify individual transponders. The following parameters are sent with this command:

  Divide ratio. Defines the frequency of transmission between the transponder and the interrogator.

  Cycles per second. Sets the data rate and the modulation format.

  TRext. Used to switch ON or OFF the preamble pilot tone.

  Sel. Selects the transponders that will respond (all of them or a selection)

  Session. Starts a session. Sessions are named S0, S1, S2, and S3.

  Target. Selects between A and B inventory flags.

  Q. Sets the number of slots for the inventory round to be used by the slot counter.

  CRC-5. 5-bit CRC calculated over the command string.

- The *query adjust* command is used to instruct the transponder to increment or decrement its slot counter value. There are two parameters sent with this command:

  Session. S0, S1, S2, or S3. This command ensures that the interrogator is talking to the correct transponder.

  Up/Dn. Instructs the transponder to count up or down.

- The *Query_Rep* command is used to instruct the transponder to decrement its slot counter value. If the resulting value is zero, it

TABLE 8.35
Example of Packet Request Structure Using the Select Command

Select Command	Target: SL	Action	EPC Memory Bank (01)	Start Address	Length (8 bits)	Mask	Trunc	CRC-16
1010	100	000	10	01111000	0001000	00000000	1	0111100000010011

transmits a random 16-bit number to the interrogator. This command only needs one parameter:

Session. S0, S1, S2, or S3 is needed to confirm the session from the inventory round.

- The *Req_RN* command is used by the interrogator to instruct the transponder to generate and return a new 16-bit random number. This new random number is called the *handle* and will be used as the identifier between the transponder and interrogator for further commands. After transmitting back this number, the transponder changes its state to open or secure depending on the status of the password. In either state, the transponder is able to accept and process new commands from the interrogator.

- The *read* command allows reading the memory in the transponder in multiples of 16-bit blocks. The *read* command requires the following parameters:

MemBank. Specifies which memory areas will be read.

WordPtr. Defines the address for the first word to be read.

WordCount. Defines the number of 16-bit words that will be read.

Handle. 16-bit identifier between transponder and interrogator.

crc-16. 16-bit CRC calculated from the first command bit to the last CRC bit.

- The *write* command allows writing a word of data to any of the memory locations in the transponder. Before this command can be executed, it is necessary to request a new 16-bit random number. The parameters used with this command are as follows:

MemBank. Specifies which memory areas will be read.

WordPtr. Defines the address for the first word to be read.

Data. 16-bit data word to be written. These data are XORed with the newly generated 16-bit random number.

Handle. 16-bit identifier between transponder and interrogator.

CRC-16. 16-bit CRC calculated on the command string.

- The *kill* command is used to disable a transponder permanently. Due to the irreversible process, the command needs some additional steps to prevent the accidental killing of other transponders in the vicinity of the interrogator. First, the interrogator requests a new 16-bit random number (RN16) from the transponder. Then, the 16 MSBs of the *kill* password are XORed by the interrogator with RN16, and the resulting bits are sent to the transponder. The interrogator requests a new RN16 that is then XORed with the 16 LSBs of the kill password. After receiving this transmission, the transponder will not respond to any request or commands from the interrogator.

- The *lock* command is used to lock individual passwords preventing reading and writing the password, to lock individual banks preventing new writing to the banks, or to make the password or memory bank permanently locked and unchangeable. The *lock* command requires the following parameters:

Payload. This is a 20-bit mask indicating the memory locations and locking actions to be performed at those locations.

Handle. 16-bit identifier between transponder and interrogator.

CRC-16. 16-bit CRC calculated on the previous data string.

Table 8.36 displays the meaning of the 20 bits that comprise the payload. For the first 10 bits (mask bits), the bit set to *0* means to keep the current settings, ignoring that particular action, while the bit set to *1* means to implement the action that is associated with the action field that is composed of the 10 last bits. The meaning of these last 10 bits (action bits) is shown in Table 8.37. Table 8.38 describes the effects of the different password and lock bits.

- The *access* command is necessary when the access password is different than zero and the transponder has been put into a secured state. To execute the access command, the interrogator requests a new RN16 from the transponders. The bits of this new RN16 number are XORed with the 16 MSBs of the password and sent to the transponder. The interrogator proceeds to request a new random number

TABLE 8.36

Payload Bits in Lock Command

1	2	3	4	5	6	7	8	9	10	11	12	13	14	15	16	17	18	19	20
Kill mask		Access mask		EPC mask		TID mask		User mask		Kill action		Access action		EPC action		TID action		User action	

TABLE 8.37

Action Bit Options

Action Bit Option	Kill Action		Access Action		EPC Action		TID Action		User Action	
	1	2	3	4	5	6	7	8	9	10
	Passwd read or write	Perm lock	Passwd read or write	Perm lock	Passwd write	Perm lock	Passwd write	Perm lock	N/A	N/A

TABLE 8.38

Effect of the Different Combinations of Password and Lock Bits

Password status	Perm lock status	Effect
Passwd Write = 0	Perm lock = 0	Memory bank writable from Open or Secure states
Passwd Write = 0	Perm lock = 1	Memory bank permanently writable from Open or Secure states. Can never be Locked
Passwd Write = 1	Perm lock = 0	Memory bank writable from Secure state but not from Open state
Passwd Write = 1	Perm lock = 1	Memory bank not writable
Passwd Read/Write = 0	Perm lock = 0	Password readable and writable from Open or Secure states
Passwd Read/Write = 0	Perm lock = 1	Password permanently readable and writable from Open or Secure states
Passwd Read/Write = 1	Perm lock = 0	Password only readable and writable from Secured state
Passwd Read/Write = 1	Perm lock = 1	Password not readable or writable from any state

from the transponder that in turn is XORed with the 16 LSBs from the password and sent to the transponder. After having transmitted successfully the second set of XORed bits, the transponder is set into the secure state.

- The *block write* command allows one to write multiple words into the EPC or reserved memory in the transponder using a single command. The parameters used in this command are as follows:

MemBank. Specifies the memory area to write.

WordPtr. Defines the address for the first word to be written.

WordCount. Defines the number of 16-bit words that will be written.

Data. Data to write.

Handle. 16-bit identifier between transponder and interrogator.

CRC-16. 16-bit CRC calculated on the data string.

- The *block erase* command allows erasing multiple words using a single command. The command uses the following parameters:

MemBank. Specifies the memory bank to be erased.

WordPtr. Defines the address for the first word to be erased.

WordCount. Defines the number of 16-bit words that will be erased.

Handle. 16-bit identifier between transponder and interrogator.

CRC-16. 16-bit CRC calculated on the data string.

TABLE 8.39

Commands Supported by LF Atmel® Transponders

Command	SOC	OpCode	Description
Read Single Block[a]	00	01	Read single 32-bit data block
Read Multiple Blocks[b]	00	01	Read multiple data blocks
Write Single Block[c]	00	01	Write a single data block
Login_Write[d]	00	0111011110	Login for write password protected access
Login_Read[e]	00	0111011010	Login for read password protected access
Get_ID[f]	00	0000	Starts a complete new anticollision loop
Get_ID (Tag ID – partial even)[g]	00	0000	Anticollision loop with partial ID. Even number of matching ID bits
Ged ID (Tag ID- partial odd)	00	001	Anticollision loop with partial ID. Odd number of matching ID bits
Select[h]	00	0000	Puts specified transponder into selected state
Select_All[i]	00	1000	Selects all transponders within range
Select_Group[j]	00	100[0]n1	Selects a specified group of transponders
Select_N_Group[k]	00	101[0]n1	Selects all transponder that are not members of the group
Reset_Selected[l]	00	11100000	Resets selected transponder to Ready state without reloading configuration register
Reset_to_Ready[m]	00	11000000	Resets all transponders in the field to Ready state. Reloads configuration register from system memory
Arm_Clear[n]	00	11001000	Arms transponder for Clear_All command
Clear_All[o]	00	01011111	Clears memory except traceability data

[a] *Read single block* command reads a complete data block (32 bits) for a transponder in the *selected* state. An optional CRC for the block address can be included in the command. If this is available, the transponder checks the CRC and aborts the operation if the command is not compatible with the received address. When the CRC check is valid or it is not included, the transponder responds with the 32-bit data followed by a CRC. This CRC code is generated over the 32-bit data as well as the memory address used. If the attempted read address is not existent or it is protected, the returned data will be a string of 1's. In order to read protected memory blocks, this command must be preceded by a successful LoginRead command. The request packet for this command requires the 6-bit block address that is transmitted with its MSB first.

[b] *Read multiple blocks* command reads an array of consecutive 32-bit data blocks from a start address in a transponder that is in the *selected* state. An optional CRC for the address can also be included with the command. When the CRC check is valid or it is not included, the transponder responds with the 32-bit data followed by a CRC. This CRC code is generated over the 32-bit data as well as the memory address used. If the attempted read address is not existent or it is protected, the returned data will be a string of 1's. In order to read protected memory blocks, this command must be preceded by a successful LoginRead command. The request packet for this command requires two parameters: 6 bits for the address of the start block, and 6 bits for the address of the end block.

[c] *Write single block* command programs a specific memory address in a transponder with a 32-bit block of data and associated lock command. To perform this command, the transponder must be in the *selected* state. If the destination block is password protected, a LoginWrite must be executed before writing. Memory blocks for which their lock bit is set to 1 cannot be written.

TABLE 8.39 (CONTINUED)

Commands Supported by LF Atmel® Transponders

Attempting to write to a locked address results in the immediate termination of the command followed by the appropriate error codes. The command includes a CRC that can be mandatory or optional depending on the status of the configuration register. After successfully receiving the write single block command, the transponder starts the sequence for programming the EEPROM. The programming process includes an automatic read verification to ensure the integrity of the data. To signal the completion of a successful programming cycle, the transponders return an SOF string. The request packet requires three parameters: the 6-bit block address, the lock bit status after programming (2 bits), and the 32-bit data to be written in memory.

d *Login_Write* command is used to release the protection against writing on all write-protected data blocks. If the password transmitted with the command is successful, the transponder will generate an SOF string to indicate the success; otherwise, the SOF will be followed by the appropriate error code. After executing this command successfully, all the write-protected memory blocks can be modified as long as the addressed memory block is not already locked. The request command requires three parameters: the string 110111, the string 10, and the 32-bit write password.

e *Login_Read* command is used to release the protection against reading on all the data blocks that are set to be read-protected. If the password transmitted with the command is successful, the transponder will generate an SOF string to indicate the success; otherwise, the SOF will be followed by the appropriate error code. After executing this command successfully, all the read-protected memory blocks can be read normally. The request packet requires three parameters: the string 110110, followed by the string 10, and followed by the 32-bit read password.

f *Get_ID* command causes all transponders in a ready state to engage into the anticollision loop. This command does not require any additional parameters.

g *Get_ID (partial)* command is used following the Get_ID command. All transponders with partial matching IDs reply with their own anticollision response by transmitting the SOF string followed by their IDs.

h The *select* command causes the addressed transponder to respond with the 16-bit CRC for its ID, followed by entering into the selected state. After issuing this command, the interrogator can communicate with the addressed transponder until it selects another one. In this case, the first transponder enters the quiet state. This parameter requests the ID of the selected transponder as parameter.

i The *Select_All* command causes all transponders in the ready state to enter into the selected state and answer with an SOF string. This command allows for global configuration of a set of transponders without having to program each one of them sequentially. This command does not require additional parameters.

j The *Select_Group* command causes all transponders in the ready state with matching partial IDs to enter the selected state and answer with the SOF string. The length of the partial ID can vary. This information is the parameter required by the command.

k The *Select_N_Group* command causes all transponders in the ready state with no matching partial IDs to enter the selected state and answer with the SOF string. The length of the partial ID can vary. This information is the parameter required by the command.

l The *Reset_Selected* command resets the selected transponders back to the ready state. After receiving the command, transponders respond with the SOF string indicating that they are able to participate in future anticollision sequences. This command does not require any additional parameters.

(Continued)

TABLE 8.39 (CONTINUED)

Commands Supported by LF Atmel® Transponders

^m The *Reset_to_Ready* command sets all the transponders in the radiofrequency field into the ready state. Depending on the configuration of the transponders, these may start transmitting data. After receiving the command, transponders respond with the SOF string indicating that they are able to participate in future anticollision sequences. This command does not require any additional parameters.

ⁿ The *Arm_Clear* command prepares transponders for the subsequent Clear_All command as long as the master key was not set to 6. If this command is followed by any command other than the Clear_All command, it becomes disabled. This command does not require additional parameters.

^o The *Clear_All* command clears all the memory blocks and their lock bits for all transponders that had been previously armed with the Arm_Clear command. The clearing of the EEPROM can be understood as a programming process of writing 0's to all memory locations, thus following the same rules as writing to memory. After the transponder memory has been successfully cleared, the transponders respond with the SOF string. If the command fails for any reason, the transponder generates the appropriate error codes. Because of its work with the Arm_Clear command to prepare the transponders for clearing, it can be seen as an additional step in avoiding accidental erasing of data in the memory of the transponders. This command requires three parameters: 011111, followed by 00, followed by 32 bits cleared to 0.

8.5 Commands Supported by the Atmel® Family of Transponders

The functions of the logic system in the Atmel® family of transponders are the initialization and reloading of configuration in the EEPROM memory control of data read and write, data transmission, command decoding, CRC check, error detection, and handling. These functions are common to all the Atmel® transponders described in this section.

8.5.1 Commands Supported by LF Atmel® Transponders

Atmel® transponders operating in the LF range are available in different memory sizes, with some of them incorporating anticollision capabilities. The commands used for these transponders, shown in Table 8.39, are divided into two parts. The first two bits, called *start of command* (SOC), are always set to 00 as they are used for autocalibration purposes. The rest of the bits, called the *OpCode*, carry the specific command information.

Table 8.40 describes the codes that can be generated as a response to an invalid command for the type of transponder receiving the command, or as a response to an invalid command sequence.

TABLE 8.40

Error Codes for LF Atmel® Transponders

Error code	Error
0111	Incorrect number of bits in command format
1110	Corrupt command encoding
0010	Attempt to write a locked block
0100	Attempt to write a protected block without previous login
1000	Command format error for Login or Write
1101	Incorrect password for Login
1011	CRC error
1010	Program 0 verification error
0110	Program 1 verification error
Other codes	Reserved

8.5.2 Commands Supported by HF Atmel® Transponders

Table 8.41 shows the commands supported by Atmel® transponders operating in the HF RFID band along with their functional classification. All these commands have a length of 12 bytes.

Table 8.42 summarizes the parameters of these data transfer commands.

8.5.3 Commands Supported by UHF Atmel® Transponders

The Atmel® transponders operating in the UHF band support two types of commands, long commands and short commands, as shown in Table 8.43.

Long commands consist of the 6-bit command and a 2-bit CRC code that the interrogator transmits to the transponder. After the transponder evaluates the CRC positively, the transponder interprets the next 8-bit stream as a parameter field. This can be followed by an address field, data field, or both depending on the command being used. The command ends with an end-of-transmission code that is built by using two end-of-file (EOF) symbols.

The *reset* command is the only command used for IC control. This is used to reset the control logic in the transponder or to reset the contents of the Status1 register. When the whole control logic is reset, the transponder does not transmit any information back to the interrogator. Group selection commands are used to address the whole memory. The page start address is fixed by the Group_ID or Group_AFI commands. Anticollision commands are used to address the whole memory. The page stat address is fixed by the Anticollision_ID command.

Short commands consist of a 4-bit command, a 2-bit modulation mode, and a 2-bit CRC code. Short commands do not have a parameter field following the actual command. The 2-bit modulation code is as follows:

TABLE 8.41

Commands for HF Atmel® Transponders

Command	Functional Classification
REQB/WUPB[a]	Anticollision
ATTRIB[b]	Anticollision
HLTB[c]	Anticollision
SlotMARKER[d]	Anticollision
READ[e]	Data Transfer
WRITE[f]	Data Transfer
LOCK[g]	Data Transfer
CHECK_PASSWORD[h]	Data Transfer
DESELECT[i]	Data Transfer
COUNT[j]	Data Transfer

[a] *REQB/WUPB* command is used by the interrogator to search for transponders in its radio-frequency field or to wake up those transponders in the halt state. The first byte in this command must be $05, and the second byte must be either $00 or $01. The third byte of the command is used to select between the REQB (search for transponders in the field) and WUPB (wake up transponders) functions. The response to this command by the transponder is a packet containing information about its identification data, application data, and other parameters of interest.

[b] *ATTRIB* command is used to select the transponders that have responded to an REQB/WUPB command. The transponder responds to ATTRIB commands with bytes that match their identification stored in memory. Those transponders that respond to this command are placed in the active state, allowing responding to data transfer commands.

[c] *HLTB* command is used to set a transponder into a halt state. In the halt state, transponders can only respond to the WUPB command.

[d] *SlotMARKER* command is used by the interrogator to query transponders for their random number status used for anticollision protocols.

[e] *READ* command causes the transponder to transmit the 64-bit data to the interrogator. A password is not required to read pages 0, 1, and 2 for these interrogators, but is necessary to access other pages. Page 3, which stores the password, cannot be read directly. The response of a valid READ command is a 12-byte data frame containing the referenced 64-bit data.

[f] *WRITE* command stores data into the 64-bit page referenced with the command. The data to write are sent immediately following the command. Those pages are read-only, are locked, or, in cases where the proper password has not been used, cannot be written. This generates the appropriate error code.

[g] *LOCK* command locks the memory address specified in the command to future changes, although pages can still be read using the appropriate password.

[h] *CHECK_PASSWORD* command is used to open a transponder to future commands. The 64-bit data embedded with the command are checked against the stored password in page 3 of the transponder. This command needs to be executed before issuing most commands. Once the transponder is open, the password can be changed by writing new data into page 3 that store the password using the WRITE command.

[i] *DESELECT* command is used to put the transponder into the halt state. This command has the same effects as the HLTB command used for anticollision functions.

[j] *COUNT* command is used to write to page 2 of the memory that contains the identification data and the count counter. The counter is a 16-bit data block that is incremented each time that this command is executed until it reaches the value $8000, after which it cannot be further incremented. At this point, page 2 becomes locked against future modifications.

TABLE 8.42

Data Transfer Parameters for Selected Commands

Command	Byte 1	Byte 2	Bytes 3–10	Bytes 11–12
READ	$4	Address	----	CRC
WRITE	$3	Address	64-bit data	CRC
LOCK	$2	----	32 first bits ignored 32 lasts bits data	CRC
CHECK_ PASSWORD	$6	----	64-bit data	CRC
DESELECT	$A	----	----	CRC
COUNT	$E	----	16 first bits ignored 48 lasts bits data	CRC

00: NRZI soft locked

01: 3Phase1

10: NRZI notch locked

11: FMO

TABLE 8.43

Commands Supported by UHF Transponders

Short / Long Command	Command Type	Command Name	Command Code	Description
Long	IC Control	Reset[a]	001010	Reset of circuit or parts of status1 register
Long	Group Selection	Group_ID	000011	Select group through ID
Long	Group Selection	Group_AFI	000001	Select group through AFI
Long	Group Selection	Group_p	010010	Group select supporting address and bit pointer
Long	Group Selection	Group_p_leeq	010110	Group select supporting address and bit pointer
Long	Group Selection	Group_p_greq	010111	Group select supporting address and bit pointer
Long	Anticollision	Anticollision_ID	000000	Anticollision targeting ID
Long	Anticollision	Anticollision_p	010011	Anticollision supporting address and bit pointer
Long	Anticollision	Anticollision_p_random	010101	Anticollision supporting address, bit pointer and random
Long	Reading	Read32[b]	000100	Read 32 bits in loop
Long	Reading	Read32c	100100	Read 32 bits in loop
Long	Reading	Read128[c]	001100	Read 128 bits in loop
Long	Reading	Read128c	100110	Read 128 bits in loop
Long	Program	Program4byte[d]	001000	Immediate 32-bit programming
Long	Program	Program4bytec	100000	Immediate 32-bit programming
Long	Program	Programmbyte[e]	011000	Up to 32-bit programming. Possible to select 8 bit portions
Short	Reading	Get_ID_page[f]	0101	Same as read128c
Short	Reading	Get_system[g]	0110	Reads system information
Short	Anticollision	Wakeup_s	0001	Answer in one of 16 slots (selected transponders)
Short	Anticollision	Wakeup_sb	0010	Answer in one of 16 slots (nonselected transponders)
Short	Anticollision	Repeat_arb	0100	Repeats long arbitration

Short	Anticollision	Slot	1000	Aloha
Short	Anticollision	Slot_selected	1001	Aloha
Short	Anticollision	Slot_not_selected	1010	Aloha
Short	Anticollision	Slot_repeat	1011	Aloha
Short	Anticollision	Skip_slot	1100	Aloha
Short	Anticollision	Slot_close	1101	Aloha
Short	Temporary	Repeat_arb		Speed up anticollision
Short	Temporary	Skip_slot		Aloha

a The *reset* command generates a complete or partial soft reset in the transponder. The reset command must be followed by the 8-bit parameter specifying the actions of the command and the 16-bit CRC. When bit 7 (MSB) of the parameter is set to 1, the transponder does not transmit a response packet.

b The *Read32* and *Read32c* commands are used to read memory. Read32 reads a 32-bit block of user memory, while Read32c reads a 32-bit block of the control memory. Other than this distinction, both commands are similar. The Read32 or Read32c commands must be followed by the 8-bit parameter, the optional address field that is only used in the long addressing mode, the CRC code, and two end-of-file (EOF) symbols. After executing the command successfully, the transponder returns the status2 register, the 32-bit data block that has been read, and the CRC code.

c The *Read128* and *Read128c* commands are also used to read memory. Read128 reads a full page of user memory, while Read128c reads a full page of the control memory. Both commands must be followed by the 8-bit parameter, the optional address field that is only used in the long addressing mode, the CRC code, and two end-of-file (EOF) symbols. After executing the command successfully, the transponder returns the status2 register, the 128-bit data block that has been read, and the CRC code.

d The *Program4byte* and *Program4bytec* commands enable the interrogator to write the user and control memory respectively. These commands can be used in long mode (addressed) or short mode (nonaddressed). After receiving the command, the interrogator must transmit the 8-bit parameter, the address (only in long mode), the 32-bit data, and the CRC code.

e The *Programmbyte* command is a long command that requires an address field. This command is used to write to a portion of the selected block. The block number and page information are transmitted in the address field.

f The *Get_ID_page* command returns the ID stored in memory. This is a short command that does not require an address or an additional parameter. After the transponder has received the command, it transmits the 8-bit Status2 register data, followed by the 128-bit ID page, finishing with the CRC code for all of the data stream.

g The *Get_system* command is also a short command. The transponder transmits the 8-bit Status2 register data, the 320-bit system memory data, followed by the CRC code.

References and Further Reading

Because radiofrequency identification (RFID) is a relatively new technology, the most current information about the different parts that compose the whole RFID system come directly from the manufacturers. For this reason, the majority of the documents that are referenced here are data sheets and application notes from the key manufacturers of RFID components. Although these references do not contain URLs because of their frequent change, interested readers should not experience any problems locating these documents in the manufacturer's websites. The books listed in this section discuss fundamental work in their subject area.

1 Books

Balanis, C. *Antenna theory: Analysis and design*, 3rd ed. Hoboken, NJ: Wiley-Interscience, 2005.

Cheng, D. *Fundamentals of engineering electromagnetics*. Reading, MA: Addison-Wesley, 1993.

Dobkin, D. M. *The RF in RFID: Passive UHF RFID in practice*. Burlington, MA: Newness, 2008.

Johnson, R. *Antenna engineering handbook*, 3rd ed. New York: McGraw-Hill Professional, 1992.

Schmitt, R. *Electromagnetics explained: A handbook for wireless/RF, EMC, and high-speed electronics*. Burlington, MA: Addison-Wesley, 2002.

Ulaby, F. *Fundamentals of applied electromagnetics*. Upper Saddle River, NJ: Addison-Wesley, 2004.

2 Technical Literature from Atmel®

2.1 Application Notes from Atmel®

ATA5590 Tag antenna connection. Document 4936B-RFID-08/07.

ATA5590 Tag antenna matching. Document Rev. 4843A-RFID-02/05.

Tag tuning/RFID. Document Rev. 2055A-RFID-07/02.

U2270B Electronic immobilizers for the automotive industry. Document Rev. 4661A-RFID-06/03.

U2270B Extended RFID reading distance by using U2270B with an additional preamplifier. Document Rev. 4592A-RFID-12/02.

2.2 Data Sheets from Atmel®

AT88RF020 13.56 Mhz, 2048-bit RFID EEPROM. Document Rev. 2010C-RFID-2/06.

AT88SC6416CRF. 13.56 MHz CryptoRF EEPROM memory 64 kbits. Document Rev. 5006CS-CRRF-12/06.

AT88SC3216CRF. 13.56 MHz CryptoRF EEPROM memory 32 kbits. Document Rev. 5005CS-CRRF-12/06.

AT88SC1616CRF. 13.56 MHz CryptoRF EEPROM memory 16 kbits. Document Rev. 5026CS-CRRF-12/06.

AT88SC0808CRF. 13.56 MHz CryptoRF EEPROM memory 8 kbits. Document Rev. 5027CS-CRRF-12/06.

AT88SC0404CRF. 13.56 MHz CryptoRF EEPROM memory 4 kbits. Document Rev. 5023CS-CRRF-12/06.

AT88SC0204CRF. 13.56 MHz CryptoRF EEPROM memory 2 kbits. Document Rev. 5022CS-CRRF-12/06.

AT88SC0104CRF. 13.56 MHz CryptoRF EEPROM memory 1 kbits. Document Rev. 5021CS-CRRF-12/06.

ATA5558 Preliminary. 1 kbit R/W IDI© with deterministic anticollision. Document Rev. 4681C-RFID-09/05.

ATA5567 Multifunctional 330-bit read/write RF identification IC. Document 4874E-RFID-10/07.

ATA5590 TAGIDU™. 1.3-kbit UHF R/W IDIC© with anti-collision function. Document 4817C-RFID-03/07.

TK5530 Read-only transponder. Document Rev. 4683D-RFID-11/05.

TK5551 Standard read/write ID transponder with anti-collision. Document 4709F-RFID-06/06.

U2270B Read/write base station. Document 4684E-RFID-02/08.

U3280M Transponder interface for microcontroller. Document 4688D-RFID-03/07.

3 Technical Literature from Microchip

3.1 Application Notes from Microchip

AN232 Low-frequency magnetic transmitter design. Document DS00232B. 2008.

AN678 RFID coil design. Document DS00678B. 1998.

AN680 Passive RFID basics. Document DS00680D. 2004.

AN680 RFID tag and COB development guide with Microchip's RFID devices. Document DS00830B. 2002.

AN707 MCRF 355/360 application note: Mode of operation and external resonant circuit. Document DS00707B. 2001.

AN710 Antenna circuit design for RFID applications. Document DS00710C. 2003.

AN752 CRC algorithm for MCRF45X read/write device. Document DS00752A. 2001.

AN759 Interface control document for the 13.56 MHz anti-collision interrogator. Document DS00759B. 2002.

AN912 Designing LF talkback for a magnetic base station. Document DS00912A. 2004.

ASK Reader reference design. Document DS5116C. 1998.

FSK Reader reference design. Document DS51137B. 1998.
FSK Anti-collision reader reference design. Document DS5167B. 1998.
PSK Reader reference design. Document DS51138B. 1998.
MCRF2XX using the microID® programmer. Document DS51139C. 2004.
MCRF355/360 Reader reference design. Document DS21311B. 2001.
MCRF45X Reference design. Document DS21654C. 2004.

3.2 Data Sheets from Microchip

MCRF200 125 kHz microID® passive RFID device. Document DS21219H. 2003.
MCRF 355/360 13.56 MHz passive RFID device with anti-collision feature. Document DS21287F. 2002.
MCRF450/451/452/455 13.56 MHz read/write passive RFID device. Document DS40232H. 2003.

4 Technical Literature from Texas Instruments

4.1 Application Reports from Texas Instruments

Antenna selection guide. Application note AN058. Document SWRA161.
Constructing a 1000 x 600 HF antenna. Document 11-08-26-007. 2003.
HF antenna cookbook. Document 11-08-26-001. 2004.
HF antenna design notes. Document 11-08-26-003. 2003.
Integrated TIRIS™ RF module TMS3705A Introduction to low frequency radar. Document SCBA020. 1999.
LF reader synchronization. Document SCBA019. 2004.

4.2 Reference Guides from Texas Instruments

Antenna reference guide. Document SCBU025. 1996.
Antenna tuning indicator. RI-ACC-ATI2. Document SCBU031. 2001.
Series 2000 reader system. 4-channel TX/RX multiplexer module for remote antenna RFM RI-MOD-Tx8A. Document 11-06021-039. 2000.
Series 2000 reader system. ASCII protocol. Document SCBU028. 2000.
Series 2000 reader system. Control modules RI-CTL-MB2A, RI-CTL-MB6A. Document SCBU024. 2000.
Series 2000 reader system. High performance reader frequency module. RI-RFM-007B. Document SCBU022. 2002.
Series 2000 reader system. High performance remote antenna-reader. Frequency module RI-RFM-008B. Antenna tuning board RI-ACC-008B. Document SCBU023. 2002.
Series 2000 reader system. Micro-reader RI-STU-MRD1. Document SCBU027. 2000.
Series 2000 reader system. Mini-RFM RI-RFM-003B. Document SCBU021. 2000.
Series 2000 reader system. Reader S251B RI-STU-251B. Document SCBU025. 2000.
Series 2000 reader system. RFM sequence control. Document 11-06021-049. 1999.

Series 2000 reader system. Standard radio frequency module RI-RFM-104B. Document 11-06-21-035. 2002.

Series 2000 reader system. TIRIS bus protocol. Document SCBU026. 2000.

Reader series 4000. S4100 Multi-function reader module RF-MGR-MNMN. Base application protocol reference guide. Document 11-01-21-700. 2003.

Reader series 4000. S4100 Multi-function reader module RF-MGR-MNMN. ISO 14443 library reference guide. Document 11-01-21-702. 2003.

Reader series 4000. S4100 Multi-function reader module RF-MGR-MNMN. ISO 15693 library reference guide. Document 11-06-21-707. 2004.

Reader series 4000. S4100 Multi-function reader module RF-MGR-MNMN. Tag-it™ library reference guide. Document 11-06-21-708. 2003.

HF reader system series 6000. S6350 Midrange reader module RI-STU-TRDC-02. Document 11-06021-700. 2002.

HF reader system series 6000. S6700 Multi-protocol transceiver IC RI-R6C-001A. Document SCBU030. 2005.

HF reader system series 6000. Gate antenna RI-ANT-T01A. Document 11-06-21-058. 2001.

Tag-it™ Transponder inlays. Document 11-09-21-055. 2001.

Tag-it™ HF-I pro transponder inlays. Document SCBU009. 2005.

Tag-it™ HF-I standard transponder inlays. Document SCBU006. 2005.

Tag-it™ HF-I plus transponder inlays. Document SCBU004. 2005.

Tag-it™ reader system series 6000. Host protocol. Document 11-04-21-001. 1999.

Tag-it™ reader system series 6000. Reader module RI-R00-3230A. Reader module with RS232 interface RI-R00-321A. Document 11-06-21-048. 2001.

Tag-it™ transponder protocol. Document SCBU032. 2000.

Texas Instruments registration and identification system. Description of multi-page, selective addressable and selective addressable (secured) transponders. Document SCBU020. 1999.

TI UHF Gen2 protocol. Document SCBU001. 2006.

TI UHF Gen2 IC antenna design. Document SCBU015. 2006.

4.3 Data Sheets from Texas Instruments

113.56 MHz encapsulated standard transponder. Document RF-HDT-DVBE-N0. 2006.

Series 2000 antennas. Document SCBS845. 2002.

Series 2000 stick antenna. Document SCBS581. 2005.

Series 2000 control module. Document SCBS845. 2003.

Series 2000 high performance remote antenna RFM and tuning module. Document SCBS850. 2001.

Series 2000 reader S251B. Document SCB852. 2001.

HF reader system series 6000 gate antenna. 2001.

S4100 multi-function reader module data sheet. 2003.

Tag-it™ HF-I transponder inlay—square. 2005.

Tag-it™ HF-I transponder inlay—large rectangle. Document SCBS820. 2005.

Tag-it™ HF-I plus transponder inlays—large rectangle. 2005.

Index